CHROMOSOME
BOTANY

CHROMOSOME BOTANY

Archana Sharma

and

Sumitra Sen

Centre of Advanced Study on Cell and Chromosome Research
Department of Botany,
University of Calcutta
Calcutta, India

CRC Press
Taylor & Francis Group
Boca Raton London New York

CRC Press is an imprint of the
Taylor & Francis Group, an **informa** business

A SCIENCE PUBLISHERS BOOK

First published 2002 by Science Publishers, Inc

Published 2018 by CRC Press
Taylor & Francis Group
6000 Broken Sound Parkway NW, Suite 300
Boca Raton, FL 33487-2742

ISBN 13: 978-1-57808-183-7 (pbk)

This book contains information obtained from authentic and highly regarded sources. While all reasonable efforts have been made to publish reliable data and information, neither the author[s] nor the publisher can accept any legal responsibility or liability for any errors or omissions that may be made. The publishers wish to make clear that any views or opinions expressed in this book by individual editors, authors or contributors are personal to them and do not necessarily reflect the views/opinions of the publishers. The information or guidance contained in this book is intended for use by medical, scientific or health-care professionals and is provided strictly as a supplement to the medical or other professional's own judgement, their knowledge of the patient's medical history, relevant manufacturer's instructions and the appropriate best practice guidelines. Because of the rapid advances in medical science, any information or advice on dosages, procedures or diagnoses should be independently verified. The reader is strongly urged to consult the relevant national drug formulary and the drug companies' and device or material manufacturers' printed instructions, and their websites, before administering or utilizing any of the drugs, devices or materials mentioned in this book. This book does not indicate whether a particular treatment is appropriate or suitable for a particular individual. Ultimately it is the sole responsibility of the medical professional to make his or her own professional judgements, so as to advise and treat patients appropriately. The authors and publishers have also attempted to trace the copyright holders of all material reproduced in this publication and apologize to copyright holders if permission to publish in this form has not been obtained. If any copyright material has not been acknowledged please write and let us know so we may rectify in any future reprint.

Visit the Taylor & Francis Web site at
http://www.taylorandfrancis.com

and the CRC Press Web site at
http://www.crcpress.com

CIP data will be provided on request.

Cover illustration is a conglomeration of images taken from Plates 9.2, 9.3, 9.4, 9.5. These plates are reproduced in this book by permission of the respective publishers from whose publications the illustrations were reproduced, namely, CRC Press, and Harwood Academic Publishers (Gordon & Breach).

Preface

The study of chromosomes is basic to genetics. The laws of heredity embodied in genes and chromosomes are undoubtedly universal for all organisms, including plants, animals, and humans. Notwithstanding this general uniformity, the chromosomes of plants, their behavioral characteristics, and role in evolution show certain distinctive features. In addition to polyploidy and the location, nature, and amount of repeat sequences, the totipotency of the plant cell enables it to have the unique potential of generating genetic diversity and chromosome variants.

These reasons led us to plan a book on plant chromosomes covering all the basics on the one hand and advances including molecular characteristics on the other.

The treatment of the subject in *Chromosome Botany* aims to present the chromosomes as a dynamic entity against the backdrop of growth, differentiation, reproduction, and evolution. Structural details, identification of gene sequences at the chromosome level, behavioral pattern, generating specific and genetic diversity in evolution, as well as the genome as affected by environmental agents, are discussed.

The objective is to provide the readers, even those uninitiated in genetics, with a concept of the totality of the chromosome as vehicle of hereditary transmission, which is the foundation of the entire discipline of molecular genetics. The authors would consider their effort successful if it proves to be useful for a beginner as well as an advanced student of plant chromosome.

The authors are grateful to successive generations of students and research scholars who had been, unwittingly or otherwise, subjected to lectures at classrooms and laboratories on these topics and also to their colleagues and families for forbearance. They are especially grateful to Professor A.K. Sharma, Founder, Centre of Advanced Study on Cell and Chromosome Research for correcting the manuscript and writing the Introduction. Thanks are also due to the Council of Scientific and Industrial Research (CSIR), University Grants Commission (UGC), Ministry of Forests and Environment

(Government of India), and Indian National Science Academy (INSA) for financial assistance.

Archana Sharma
Sumitra Sen
Centre of Advanced Study on Cell and Chromosome Research
Department of Botany
University of Calcutta
35 Ballygunge Circular Road
Calcutta 700 019
India

Plant Chromosomes Today: An Introduction

AK Sharma

The study of chromosomes in plants has become a completely synthetic discipline depending to a large measure on biophysical and biochemical tools for clarification of this material basis of heredity. At the molecular level, it represents a giant complex molecule made up of smaller but less complex molecules, the *genes*, which are arranged in a linear order in the nucleoprotein skeleton. The study has gradually been shifted to a chemical and molecular level from a purely cytogenetic standpoint. Such an analysis has given an understanding of the pattern of organization even at the submicroscopic stage. Techniques of high resolution autoradiography and electron microscopy are responsible for resolving the chromosome structure as made up of fibrils 20–30 Å in thickness, folded several times to yield a fiber 100 Å thick. It is a long chain nucleoprotein fiber in which condensed and decondensed segments alternate, condensed segments representing an aggregate of beads of nucleosomes made up of histone octamers and a surrounding DNA fiber. Pulse labelling autoradiography has further revealed that chromosome structure is made up of many replicons or replicating units, which divide as discrete units, instead of the divisions starting from one end and moving to the other. The structure is linear, multirepliconic, and uninemic.

One of the landmark developments in chromosome structure is the discovery of repeated DNA sequences. In higher organisms, in the DNA complement, multiple copies of similar sequences are present, which form the bulk of the DNA. In the linear structure of the chromosome, such highly homogenous repeats are generally located at one locus, minor and moderate repeats being interspersed throughout the unique sequences.

Chromosomes in general and plant chromosomes in particular contain a high percentage of such repeat sequences, that is similar sequences, replicated several times.

They may be major, moderate or minor, depending on the number of copies of similar sequences and the extent of homogeneity. Such repeated sequences are found within and between the genes, and form a major component of chromosome structure. Moreover, similar types of repeats may be located at one locus or may remain dispersed in several loci in the same chromosome. Their presence in different loci suggests repeated insertions.

In several plant species such as barley and rye, the repeat value may reach up to 90–95%, and only the remaining 10% of DNA constitutes genes controlling all characters.

Another signal development has been the discovery of mobile sequences, which have added a new dimension to chromosome structure. They contain a high degree of repeats and are located in different organisms, playing an important part in their evolution. This property of movement leads to insertion of repeats at different loci, both between and within the chromosomes. These sequences are now used in gene transfer, because of their property of mobility, initially observed by McClintock in maize.

Over and above these discoveries, two other important findings with regard to functional segments, which represent landmark events, need special mention. Telomeres or chromosome ends have now been shown to have a special type of structure with repeat overhang and to have a novel method of origin. They are synthesized by telomerase, a ribozyme in which instead of protein, RNA forms a core and acts as a template for synthesis of telomeric DNA. The telomeric ends also undergo regular shortening with ageing and inhibition of shortening may lead to excessive proliferation. The other functional segment – the centromere – too, as recent studies show, not only acts as an attachment point of the spindle but is also the site of action of motor protein.

Simultaneous to such advances, aided by biochemical and molecular methods, the refinements in cytological and cytochemical techniques have led to the clarification of chromosome complements from any organ of the plant system.

Technical innovations for study of the structure and behavior of chromosome have led to unravelling of some of the unexplored basics of chromosomal control of growth, differentiation, and reproduction in the plant system. The dynamic role of repeat sequences in the control of nonspecific functions is gradually being realized. The behavioral differences of chromosomes, during organogenesis, involving DNA–protein interaction along with endoreplication, are gradually coming into focus. Even control of chromosomal aberrations in somatic tissue of asexual species for generating diversity is providing facts basic to speciation. The dynamic behavior of chromosomes along with its repeat sequences is being increasingly realized in terms of growth, differentiation, and reproduction.

In fact, at the level of chromosomes, behavioral differences, gene mobility, sequence changes and repeat dynamism contribute to enrichment of genetic diversity.

In the study of phylogeny and evolution, the contributions of repeat sequences in conservation of gene order are becoming apparent. A comparison of the genomic map of several grass genomes reveals remarkable similarities in the order of gene sequences

– a phenomenon termed *synteny*. From a simple genome, it is claimed that the biodiversity in evolution has been aided to a great extent through expansion of repeated sequences, duplication, and polyploidy.

Along with the resolution of structure and behaviour of chromosomes, there have been remarkable developments in the analysis, manipulation, and engineering of chromosome segments. The use of probes, laser dissection, and polymerase chain reaction (PCR) have indeed revolutionized the study of structure and function of chromosome sequences to a significant extent.

The use of probes and laser in identifying gene loci in chromosomes, as based on the principle of molecular hybridization, is a major development. In the identification of chromosome segments, molecular hybridization *in situ* has been much refined in recent years, substituting isotopic probes with non-isotopic ones. This technique has been modified with the use of multiple probes, tagged with compounds of different colors, so that a number of genes can be simultaneously located in the chromosomes. They can be detected through differential fluorescence at different chromosome loci. The entire procedure of locating a large number of genes simultaneously in the chromosome with the aid of different fluorescent colors is otherwise termed as *"chromosome painting"* at the molecular level. This method permits even detection of single copy sequences of genes in chromosomes.

The scope of *in situ* localization of functional gene sequences at the chromosomal level with different probes has been further widened with mapping of the *Arabidopsis* genome and availability of a wide spectrum of probes.

The genes, once localized in the chromosomes, can be isolated for analysis and manipulation. At the chromosome level, isolation and separation of chromosome fragments have been much simplified in recent years by laser dissection of chromosome under microscope utilizing a micromanipulator as well. The manipulation of minute gene sequences at the molecular level under the microscope, a remarkable feat of engineering, has become a reality because of laser surgery.

For further analysis and convenience in manipulation with large amount of gene material, DNA sequences can be subjected to PCR, which enables a billion-fold amplification of short sequences within a few hours. Thus use of specific probes for identification, laser beam for dissection, and *in situ* PCR for amplification have revolutionized the chromosome study at the molecular level. Visual manipulation of genes under the microscope permits chromosome engineering with predictable accuracy.

Advances in chromosome research have undoubtedly led to an increased understanding of different aspects of genomics and amplification of gene loci in situ – their dissection, processing, and even transposition at chromosome site have all become realities. Strategies of analysis, manipulation, and engineering owe undoubtedly to innovations of novel technologies. But the crucial factor controlling the success of all these methods is the preparation of the basic material – the chromosome–with clarified details at the cellular level. All the analysis and manipulations can be carried out with accuracy provided chromosome complements are obtained that are distinct under the microscope. To achieve this objective of securing excellent chromosome preparations, basic to all approaches on chromosome research, some simple techniques for the study of mitotic and meiotic chromosomes from different plant organs are given in the Appendix. The simplicity of these methods may generate enthusiasm to one uninitiated in chromosome research.

Centre of Advanced Study on Cell and Chromosome Research
Department of Botany
University of Calcutta

CONTENTS

CHAPTER 1

Chromosome: Structure and Components

Introduction

The term "chromosome", coined by Baranetsky, is the principal component in the nucleus of higher plants, and is responsible for maintenance of hereditary stability. The chromosomes per se, containing the genes, form the material basis of heredity, the genes being arranged in a linear order. The chromosome complement of each cell contains the total gene content and constitutes the genome of the individual. The behavior of chromosomes follows a distinct pattern in the body cell or soma as well as in the germ cell line. The chromosome itself divides longitudinally into two halves, each half going to a daughter cell. Such equational separation of chromosomes as a whole, embodied in cell behavior termed *mitosis*, permits each and every cell of an organism to have same chromosome component and number. During the formation of gametes, responsible for sexual reproduction, the chromosomes adopt a different behavior. This behavior, otherwise known as *meiosis*, involves the coming together of the maternal and paternal chromosomes, which

pair, forming bivalents, cross over and exchange segments followed by the two different chromosomes separating to two different poles. The daughter nuclei following such division contain half the number of chromosomes. Such nuclei with reduced number of chromosomes occurring on both the male and female side may go on multiplying to form gametes. During reproduction, the male and female gametes unite to form the zygote. After fertilization, the original number is restored. The continued equational division of zygote subsequent to embryo leads to the development of the adult individual, where all cells theoretically contain the same basic chromosome complement.

The process of mitosis and meiosis, i.e. equational and reductional division, ensures that every cell of an individual contains not only the same number of chromosomes but also all the genes of the complements.

Reductional separation leads to halving of the gene complements in the gamete. The fusion of the two gametes and subsequent development of fused cell restores not only the chromosome complements but also the total gene dosage. Exchange of segments occurs during crossing over – a phase of reduction division. In general, this type of behavior, universal for higher plants and animals, ensures uniformity and stability of the gene complement, in which chances of variability are inbuilt through the recombination mechanism and gene alteration.

In the basic chromosome structure, the chromonema thread is seen to have dense thickened areas with thinner regions in between, giving the appearance of a string of beads. The microscopic beaded regions are known as *chromomeres* and their position, and size as well, are constant to a certain extent.

Chromosomes at the cytological level have been regarded as composed of units of replication, transcription, mutation, and recombination.

Structural Details and Chemical Nature

In recent years, the structure and behavior of chromosomes have been analyzed in detail, aided by gradual advances in techniques. An understanding of the chromosome structure has been achieved at structural, ultrastructural, and molecular levels. The structural details are analyzed through different approaches, and these methods, combined with cytogenetic analysis, have helped in a better understanding of the basic structural pattern of chromosomes.

The eukaryotic chromosome at the molecular level represents a single long DNA molecule or several small DNA molecules forming a continuum, arranged end-to-end as several replicons or replicating units. Serial arrangement of DNA

molecules also involves the assumption that linker sequences are present in between the different replicons or replicating units.

The chromosomes of higher organisms at the ultrastructural level have been shown to be made up of fibrils, approximately 20–30 Å in diameter, folded several times to yield a diameter of 100 Å thickness. It represents a continuous DNA–protein fiber in which condensed and decondensed segments alternate. A single condensed segment of a chromosome or chromomere of higher organisms is often comparable to the entire DNA skeleton of a microbe.

Uninemy

The earlier controversies with regard to strandedness of chromosomes have been fully resolved, and their uninemic nature has been confirmed.

The experiment of Taylor et al. (1957) using autoradiography had clearly indicated the uninemy of chromosome structure. This experiment involved treating dividing root tips of *Vicia faba* in nutrient solution with tritiated thymidine. Following replication, each daughter cell had each bipartite chromosome with one chromatid showing radioactive spots, indicating that each double helix in the chromatid had one parental strand and one new strand incorporating H^3-thymidine. In subsequent replication, 50% of the chromosomes did not show radioactive spots. With subsequent divisions, lesser and lesser number of cells showed radioactive spots (Fig. 1.1). It has been claimed further that the DNA molecule is present along the entire length of the chromosome, of course associated with histones and other

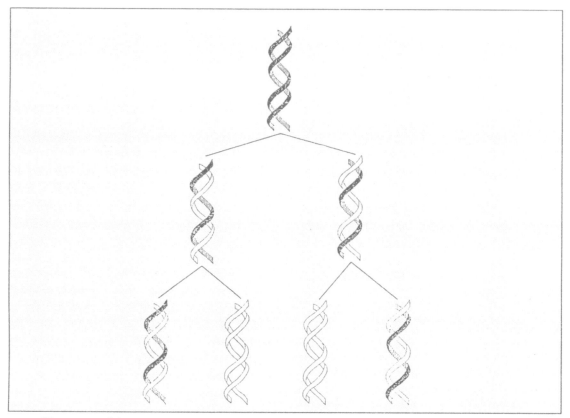

Figure 1.1 Schematic representation of the semiconservative mechanism of DNA replication.

proteins. The molecule runs from telomere to telomere uninterrupted through the centromeric region. Although the amount of DNA per chromosome is very high, reaching the value of 10^{10}–10^{11} daltons or even more in a single double-stranded DNA molecule, almost 2 m of this DNA can be confined within a nucleus less than 10 μm in diameter.

Despite the presence of both histone and nonhistone proteins in chromosomes, DNA is responsible for maintaining structural integrity of the chromosomes. It is proved by the fact that treatment with DNAse alone can destroy the structural integrity of the chromosomes (Prescott, 1970).

Replicons

In this uninemic skeleton, chromosome structure represents a linear array of potentially independent replicons. The fact that the eukaryotic chromosome is made up of several multiple discrete units for replication has been specially demonstrated through DNA fiber autoradiography in cultured cells of mammalian tissue. The patterns of linear array of silver grains with decreasing density at both ends have been taken to suggest that replication is initiated at a central region of each unit and the chain elongation is bidirectional. As such, the chromosome structure is regarded as uninemic and multirepliconic in constitution. However, theories have been proposed regarding the presence or absence of linkers in replicons. Along the DNA fiber, the initiation of replication occurs at multiple internal sites. Taylor (1984) proposed a model presuming that the functional replication may be of the size of 100–300 kilobases, as indicated by autoradiography. A subset of the dispersed repeats may be responsible for suborigins in replicon at early development as well as during differentiation. Many of the suborigins may serve as sites for cessation of replication and initiation of transcription. The replication units represent a cluster of genes including some flanking regions.

Linkers

The presence of protein, RNA, and the divalent cations Ca^{++} and Mg^{++} indicates that these might serve as linkers. To test these hypotheses, enzyme digestion tests were performed and it was seen that neither protease nor nuclease digestion could disrupt the chromosome structure. Similarly, removal of cations through chelation, also left the structural integrity of the chromosomes undisturbed. These evidences show that the replicons in the DNA molecule are not attached to each other by means of any such linkers (Prescott, 1970). Linkers between the replicons may exist between two operons, i.e. through nonsense triplets or spacers.

The advantages of the replicon model are manifold. It provides a suitable base for independent replication of different chromosome segments. Thus, early and late replication can be easily accounted for. Similarly, the termination and initiation points of replicons may suggest loci for genetic crossing over. Translocation of segments, even reciprocal and at an intercalary level, can be explained through such a mechanism.

Heterochromatin and Euchromatin

Chromosomes are composed principally of two complexes – *heterochromatin* and *euchromatin*. The structural and functional differentiation of chromosomes into heterochromatin and euchromatin, spindle organizing regions, centromeres, chromosome ends or telomeres, and nucleolus organizing regions are well documented. Euchromatin constitutes the principal functional regions of chromosomes. Heterochromatin, on the other hand, is chiefly present on both sides of the centromere, at the telomeres, secondary constrictions, and also in a number of cases at the intercalary segments.

The term "heterochromatin" was originally derived from "heterochromosome" or sex chromosome, differentiating it from autosomes, by Heitz. Since sex chromosomes differ from rest of the chromosomes in their staining behavior, the chromosome segments showing a staining cycle similar to that of sex chromosomes were called *heterochromatic*. However, Darlington attributed to heterochromatin, the property of *allocycly*. It implies positive staining in chromosomes or heteropycnosis during interphase and reverse behavior in metaphase. However, in a number of cases, chromosomes may be apparently condensed throughout divisional cycle or decondensed throughout as in secondary constriction region. As such, all these segments were clubbed under the category "heterochromatin," the common property being a staining behavior different from rest of the chromosome segments.

It was also presumed that genetically heterochromatin, though representing condensed state of chromatin, may be made up of genes having small, similar, and supplementary effects, as embodied in the concept of polygenes by Mather (1944). In later years, the discovery of repeated DNA sequences and their distribution in the chromosome indicated that heterochromatic segments are made up of repeated DNA, both GC/AT rich, the number and condensed state varying in different segments of chromosomes.

The term "heterochromatin" covers a rather heterogenous assemblage of chromatin represented by diverse characteristics in various species. In view of its repeat DNA constitution, *nucleotypic effects* have been attributed to heterochromatin. As regards function, variations on certain aspects of chromosomal metabolism, cytoplasmic synthesis, and nonspecific functions at the chromosomal level all categorized as "nucleotypic" are claimed to be regulated by these specialized segments, possibly influencing protein synthesis.

According to Vanderlyn (1949), any segment displaying properties different from those of euchromatin should be termed "heterochromatin." This statement does not necessarily imply that all heterochromatic regions, including those of primary and secondary regions, are fundamentally alike. The general terms applied to heterochromatic regions earlier were *constitutive* or inherited and *facultative* or appearing during development.

Heterochromatic property may be manifested during development, the best example of facultative heterochromatin being provided by the *mealy bug*, where one set becomes heterochromatized during development and the other remains euchromatic. In sex chromosomes of mammals, due to dosage compensation, a single X remains active, the other becoming inactive or heterochromatized. Following cold treatment, certain regions on chromosomes of plants often show negative staining as in species of *Fritillaria, Trillium,* and *Paris.* Such areas, termed by Darlington *as nucleic acid starved areas* arising out of environmental changes, come also under the category of facultative heterochromatin.

Constitutive heterochromatin, on the other hand, remains stable in the chromosome and appears condensed during interphase as in prochromosomes in a large number of plant species such as *Vicia faba.* It is also often characterized by late or early replication of DNA, different from that of euchromatin, as in case of sex chromosomes. The best example of constitutive heterochromatin is provided by segments on both sides of the centromere, which remain as a single condensed block during metaphase, being designated as *prochromosomes.*

In addition to the centromere, constitutive heterochromatin is also located at the telomeric regions, nucleolar organizing regions, as well as intercalary position in certain organisms. The entire chromosome may also be heterochromatic, such as sex chromosome in some plants, as well as supernumerary or accessory chromosomes. At the entire chromosome level, B chromosomes or accessory chromosomes are mostly heterochromatic in nature. The number of such B chromosomes in a population may or may not vary. As compared to *Allium stracheyii,* where individuals differ in the number of B chromosomes, absolute constancy of such accessory chromosomes in all individuals of a population has been recorded in *Tradescantia virginiana*.

Variability of heterochromatin and its significance

Polymorphisms and genetic diversity

Constitutive heterochromatin shows variations both at the population as well as at the species level. Polymorphisms in chromosome structure at an intraspecific level have been recorded in several species of plants (Sharma, 1974). Such polymorphism may involve difference in nature and extent of constitutive heterochromatin in different individuals. This difference has been observed in characteristically stained bands appearing as Giemsa-stained regions. Such polymorphism in bands within homologous chromosomes has also been recorded, suggesting genetic diversity at the intraspecific level.

Within a genus, different species may show variable pattern of constitutive heterochromatin. Such variations have helped in tracing species relationship, evolution of karyotype, and role of structural alterations in evolution (Sharma, 1985).

Repeated DNA sequences, as mentioned earlier, are present in heterochromatin as large segments representing highly homogenous sequences. Heterochromatic regions in cereals often show differences in tandem array of repeated sequences, conferring variability of heterochromatin in different genotypes. Species-specific highly repeated sequences have been noted in heterochromatin of *Alstroemeria*

aurea. The ribosomal RNA genes, viz. 18S, 26S, and 5S, are organized in high-copy tandem repeats and also show heterochromatic structure. Similarly, in pachytene chromosomes, rRNA genes have been located in tomato. In plant systems, variability of heterochromatin has been amply recorded as a criterion of genetic diversity.

In view of the presence of wide natural and induced variability in certain blocks of constitutive heterochromatin, it is presumed that a genome has a certain amount of tolerance so that parts of it can be lost or gained without any detectable phenotypic alterations. Also, more variation can be induced and be tolerated by cells. Therefore, studies on the variability in constitutive heterochromatin, coupled with its constitutive repetitive DNA in different cell populations of a species, may determine the tolerance of the genome for such variants and limits of this tolerance for more variability. These regions of constitutive heterochromatin may be interspersed within genes for essential sequences.

Role in amplification

Moreover, polymorphism in chromosome structure contributed by heterochromatic, segments shows quantitative DNA variations during organogenesis due to differential amplification of DNA. Such ontogenetic variations, arising out of differential amplification, have been recorded in several plants (vide Lavania, 1999; Nagl, 1978).

Researches have also revealed that chromosome constituents change their pattern during different phases of development, maintaining of course, a basic genetic make-up responsible for hereditary stability (Sharma, 1974, 1985). Chromosomes from such tissues where polytenic chromosomes are found (e.g. suspensor cells in plants or in tissues where the cell behaves endomitotically at the time of differentiation of organs), the distribution and amount of heterochromatin may be involved in the expression of character and organogenesis.

Role of recombination

In *Nicotiana* hybrids, exhibiting hybrid induced changes in chromosome morphology, evidences suggest that the ability to produce "megachromosomes" in *N. tabacum* (Gerstel and Burns, 1976) is the general property of heterochromatin expressed in an alien background.

Constitutive heterochromatin may help to recognize homologous pairing. Differences in heterochromatin content and C-banding pattern between some primitive and present-day races of *Zea mays* have been noted. Significant polymorphism for heterochromatin patterns and quantity in all chromosomes in a population of *Scilla sibirica* is also on record. Interracial hybridization may also help to generate high heterozygosity for banding patterns in relation to heterochromatin, which can be utilized for homologue recognition and chiasma terminalization.

In *Triticale*, meiotic pairing is significantly increased, thereby improving seed fertility, provided selection of the parents before the production of hybrid material of *Triticale* is made, for reduced quantity of C-band heterochromatin. Thus, the variation in C-banded heterochromatin may be a significant factor in recombination.

Role in cell cycle

The influence of heterochromatin in shortening duration of the cell cycle has been suggested, but data is limited to only a selected number of plants. The involvement of heterochromatin, repetitive DNA, and total DNA content in chromosome evolution toward control of cell cycle duration and cell growth is significant.

It has been suggested that during evolution of species, usually the change in total DNA content is, to a great extent, confined to its repetitive DNA sequences. Constitutive heterochromatin, which reflects principally the repetitive DNA fraction, possibly plays an important role in species evolution.

Recent evidences suggest that Sat DNA blocks and heterochromatin surrounding centromeres are factors influencing sister-chromatid cohesion.

Gene silencing

Lately, heterochromatin has been discussed in relation to gene silencing. It has been suggested that heterochromatin is an important constituent of the genome, though its structural and functional properties are still debated. DNA associated with heterochromatin is mostly, not exclusively, noncoding repetitive DNA, including satellite transposable DNA elements.

The term "silencing", initially used for transcriptional inactivation of certain specific gene loci in yeast, *Saccharomyces cerevisieae* (vide Henning, 1999) has later been found to occur in other genomic regions, including centromeres and telomeres. Silencing, which often follows gene duplication, may result out of direct interaction with centromeric heterochromatin. In yeast telomeres, protein complexes responsible for silencing effects, do not necessarily react only with DNA. The silencing complex involving heterochromatin can extend considerably beyond the limits of telomeric DNA.

Structural properties of heterochromatin can be extended to other chromatin regions coming in contact with inactivated heterochromatization. Genes adjacent to the telomeric region may become a part of chromatin associated with the telomere and become heterochromatic and inactive. It is likely that a particular DNA sequence may be involved in regulation of inactivation by chromatin content (Fanti et al., 1998).

Conclusion

No specific functions have been attributed to heterochromatin. Its variations at different taxonomic level make it suitable for studying biodiversity, and genotypes differ in the nature and topography of heterochromatin. The role of heterochromatin in the control of recombination is well established. The adaptational role of these segments regulating adaptive functions is evident through the behavior of certain B chromosomes, which are cytological embodiment of heterochromatin. Evidence of gene silencing is an indication of its role in gene expression. The importance of heterochromatin is indicated in adaptation and also in its involvement in the diminution of chromosome size with increasing dosage of polyploidy. Its adaptive role is indicated by involvement in gene dosage in polyploids.

Centromere

The *centromere* or the *kinetochore* is a specialized region of the chromosome associated with its movement along the spindle to the two poles during anaphase. It may be *localized* on a particular part of the chromosome – the primary constriction region – or it may be *diffuse*, in which the entire chromosome expresses the property of spindle attachment, showing parallel movement on the spindle as in species of *Luzula*. The centromere forms the final point of adhesion for the sister chromatid during metaphase to anaphase transition, attachment point for mitotic spindle fibers, as well as a site for motor protein to mediate chromosome movement (Sunkel and Coelho, 1995).

The localized centromere appears basically as a small destained gap during metaphase. This appearance is due to the presence of two blocks of heterochromatin at the two sides of the centromeric apparatus, which exhibit allocycly. They are destained at metaphase and stained at interphase to form the brightly stained bodies or prochromosomes in the interphase nucleus.

Based on the relative positions of the centromere, chromosomes are described as *metacentric, submetacentric, subtelocentric,* and *telocentric*.

During normal division, centromeric chromomeres and threads divide lengthwise so that an identical replica is transmitted to each daughter chromosome. However, under certain conditions, transverse division has been observed, referred to as *misdivision* or bursting of the centromere (Darlington, 1939). The resultant daughter chromosomes have terminal centromeres.

The *diffuse* or polycentric centromeres do not show any single attachment point. The diffuse nature of the centromeres of *Luzula* has also been confirmed through irradiation of chromosomes, where the broken fragments behave as chromosomes with centromeres. As such, in this genus, increase in chromosome number arises through fragmentation. This has also been confirmed through the measurement of total chromatin, which is identical in 2*n* = 6 in *L. purpurea* to 2*n* = 24 chromosomes in various other species (vide Camara, 1957; Nordenskiold, 1957).

Electron microscopic studies have given further details of the structure of the localized centromere. Microtubules are attached to the kinetochores, which are trilaminar plate structures on either side of the centromeric constriction. Ultrastructural analysis further shows that the two chromatids of a metaphase chromosome are held together at the centromere by two hemispheric matrices. The two arms of each chromatid are interconnected across the matrix by chromonemata (50 nm in diameter), that are continuous with the chromonemata of the chromatids. In the area of attachment to the matrix, a swelling of spindle fiber bundle is seen, called the *spindle spherule*.

Various types of connections are seen between microtubules and chromatin in plants and animals. In certain higher plants such as in species of *Lilium* and *Tradescantia*, the microtubules end in a spherical mass of fibrils, less dense than the rest of the chromosome, embedded in a cup-shaped indentation of the chromatin. In a moss, *Mnium*, the structure is very complex. The microtubules end in a band or disc 250-Å thick, separated from the chromatin by a less dense area. The dense plate seems to contain two or more 250-Å fibers that often appear double and may be continuous with chromatin on either side of the kinetochore. These fibers are surrounded by less dense fibrous material. The kinetochore has been compared with

a lampbrush chromosome loop, suggesting that it is a specialized active region.

Long blocks of repetitive sequences have been located around the centromeres of several species. In general, the centromeric DNA–protein complex is resistant to nuclease digestion in vitro. Centromeres have been shown to suppress meiotic recombination in some systems. However, in yeast strains, nonreciprocal recombination in mitosis was seen to involve the centromeric region.

Centromere: chemical nature

Plant kinetochore contains protein accumulating at the mitotic centromere in addition to those existing during interphase (Houben et al., 1996).

Many plants have clusters of highly repetitive sequences at or near their centromeres. Two different repetitive sequences almost the length of one nucleosome are present in *Brassica napus* (4*n*) and diploid ancestral species *B. campestris* and *B. oleracea*. These sequences can be used as markers.

In centromeres of monocotyledonous plants, two different conserved repeats have been found. They have been conserved in most grasses, namely rice, maize, wheat, for nearly 60 million years. These elements often show homology to the CENP-B box – a conserved 17-bp motif in humans or CDE II in yeast, bound to centromere protein B, localized in the centromere/kinetochore complex. The centromeric protein has been localized using indirect immunofluorescence with CREST antisera-anticentromere antibodies (Hadlaczky et al., 1986). Centromere of *Saccharomyces cerevisieae* consists of 3–4 conserved centromeric DNA elements

(CDEs) of about 250 bp, whereas in the fission yeast, *Schizosaccharomyces pombe*, the essential centromeric DNA consists of 44–100 kb with a 4–7 kb central core. They are species-specific and become nonfunctional in other species.

In plants, a very few cases are known of tandem repeats being specifically located at or around the centromere.

Different types of proteins too have been localized in the centromere, using auto-antibodies for autoimmune patients, and the role of nonhistone proteins in centromere structure is also distinct. The centromere-associated proteins, detectable by individual sera, show species-specific differences in their electrophoretic mobility.

Dynein, a motor protein, generates force on cytoplasmic microtubules and plays a role in spindle formation and cell division. The function of the motor is dependant on the direction of its interaction with microtubules, which are inherently polar structures. In the spindle, minus ends of microtubules are at the spindle pole and plus ends are attached to the chromosome at the kinetochore (Sawin and Mitchison, 1990).

Telomere

Telomeres are nucleoprotein structures and occur at the ends of the chromosomes. They perform several vital functions, including end protection. Telomeres enable the chromosomes to distinguish between normal chromosome ends and breaks; so that the cycle can be delayed to repair chromosome breaks. Protection of the chromosome ends is mediated through a special mechanism involving *telomerase* – a *reverse transcriptase*. This compensates for the inability of DNA polymerase to replicate chromosome completely.

Telomeric DNA, comprising the extreme molecular ends of chromosomes, consists of simple tandemly repeated sequences characterized by clusters of G residues in one strand. An overall asymmetric strand composition results in G-rich and complementary C-rich strands. The 3′ end of each strand of the duplex chromosomal DNA molecule is the G-rich telomeric strand and it forms a 3′ terminal overhang 12 to 16 nucleotides in length protruding from the duplex (Blackburn, 1990) (Fig. 1.2). Each species has a characteristic telomeric repeat sequence. Limited sequence variations are found in some species. However, widely divergent species can have the same telomeric repeat unit as well. For example, 5′-AGGGTT-3′ is the telomeric repeated sequence, both of a cellular slime mold and man. Several *Allium* species, on the other hand, lack consensus repeat arrays and instead contain complex arrays of 375-bp repeat at the chromosome termini (Fuchs and Schubert, 1996). This is possibly due to the recombination mechanism, leading to an increase in base pairs and loss of telomerase (vide Gill and Friebe, 1998).

Polymerization of TTGGGG is achieved by adding into the telomeric oligonucleotide primer at the 3′ end of a G-rich strand, independent of an exogenously added nucleic acid template. The enzyme requires a DNA primer. It can use the G-rich strand telomeric sequences from all eukaryotes tested, but not random sequence DNA oligonucleotides. Each telomerase synthesizes its species-specific G-rich strand sequence.

The existence of telomeres can explain many properties of chromosomes in vivo.

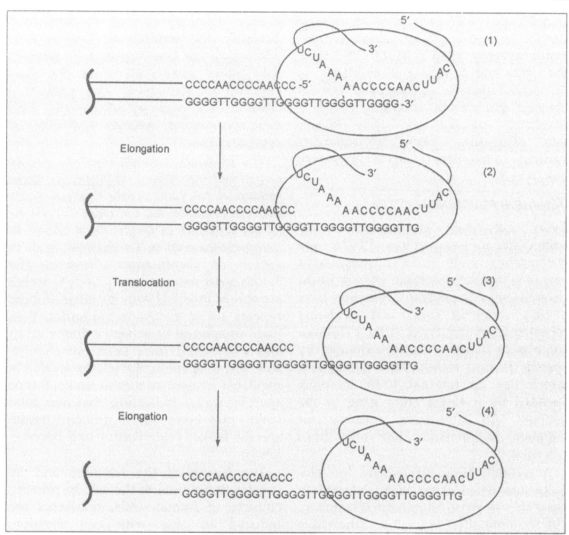

Figure 1.2   Model of elongation of telomeres by telomerase.
(1) After recognition of the TTGGGG strand by telomerase at 3' most nucleotides are hybridized to the CAACCCCAA sequence in the RNA. (2) The sequence TTG is then added one nucleotide at a time. (3) Repositioning of 3' end of TTGGGG strand leads to the 3 most TTG nucleotides undergoing hybridization to the RNA component of telomerase. (4) Elongation and copying of the template sequence to complete TTGGGGTTG. *From Carol W. Greider and Elizabeth H. Blackburn. Reprinted by permission from Nature 337: 331–336 (1989), Macmillan Publishers Ltd.*

Telomeres of one species can stabilize linear DNA molecules or chromosomes in another species, even though the two organisms have different telomeric DNA sequences. In vitro, all telomerases require a minimum length (10–12 nucleotides) of G-rich strand telomeric DNA, similar to the length of the 3' overhang of telomeric DNA, for high-affinity recognition as a primer in vitro (Blackburn, 1990, 1991).

Replication of chromosomal DNA takes place by semiconservative DNA replication. Normal DNA replication mechanism cannot lead to complete telomere replication as they yield a blunt DNA end, the other end having a 3' overhang. A "C"-strand-specific exonuclease activity converts blunt DNA ends into G-rich 3' overhangs in the cell cycle (Sawin and Mitchison, 1990). Considerable variation in telomere length is common in eukaryotes.

Telomere maintenance

Every eukaryotic chromosome must compensate for terminal loss of DNA from chromosome ends as DNA polymerase cannot completely replicate ends of linear chromosomes. In general, eukaryotes have a long array of short and tandemly repeated DNA sequences at their chromosome ends. These ends are extended by specific reverse transcriptase-telomerase, which has an internal RNA template encoded by a single copy gene in the genome. An alternative pathway for telomere maintenance has also been recorded.

A recombination mechanism for telomere maintenance has been detected in yeast also. In yeast, telomerase is common, but recombination is used as alternative pathway, telomeres being maintained about 300 bp per tip. Several genes have been identified for maintenance of telomere length. Mutations in some of these genes cause a change in the equilibrium of chromosome length. Mutants for genes responsible for activity of telomerase gradually lose telomeric sequences. Ultimately, there is chromosome loss as the telomere repeat falls below a minimum requirement. However, in case of telo-

merase inactivity or loss, recombination pathway for maintaining chromosome end length has been reported. In general, the higher eukaryotes have the dual capacity of elongating and protecting chromosome ends by telomerase, TBD (telomere bound) protein and unequal recombination.

The telomeric repeats are very simple sequences but minor changes in these sequences can have drastic changes in the cell. For example, the sequence G_4T_4/C_4A_1 of the telomere of *Oxytricha*, if added to chromosome ends in *Tetrahymena*, leads to failure of chromosome division. The *Tetrahymena* repeats are G_4T_2/C_4A_2 which are almost identical with the other. If these repeats act as a nonspecific buffer, then there should not have been a drastic effect and they could have been interchangeable. The total amount of telomeric DNA is regulated by species-specific and cell type specific ways, indicating that the total length may be related to different tissue-specific factors (vide Pardue and Baryshe, 1999).

The length of the telomere and its regulation are vital to the cell. In primary cultures of human cells, telomeres are reduced in size with cell division, becoming shorter in aged as compared with younger men. With immortalization and stabilization of culture, telomeres lengthen and telomerase activity is detected. Telomerase is very active in human tumors. Failure to find telomerase activity in normal somatic cells, may also indicate that activity of telomerase and extension of telomere end after embryogenesis (Harley et al., 1994 – vide Pardue and Baryshe, 1999).

Molecular clock

It has been suggested that the model of telomeres serves as a clock controlling life span (Kipling and Fragsher, 1999). Interstitial telomeres are also found in certain cases which arise often out of fusion of two subtelomeric chromosomes. Such telomeres are also capable of inactivating one of the telomeres in a dicentric chromosome, rendering the latter functional. The telomeric DNA in eukaryotic cell is progressively lost in successive generations on account of both incomplete replication and strand specific exonuclease activity. In case the loss is not mitigated, the telomere continues to shorten with generational doubling (Slijepcevic, 1998).

In absence of telomerase activity, chromosomes shorten at each division and active genes are exposed to cell death. Immortality is achieved by reactivating telomerase, which prevents terminal erosion. The power of telomerase to immortalize cell culture has been demonstrated in human cells. Introduction of telomerase in normal human cells leads to extension of life lines for an indefinite period (Bodner et al., 1998). Studies suggest that normal human somatic cells do not lack telomerase, but the enzyme activity is tightly growth regulated. It is visible only in actively growing cells. The complete lack of telomerase activity by knock-out experiments shows disastrous effect only after a few cell generations (Lee et al., 1998).

The simple telomeric DNA sequences are associated with complex proteins. These proteins often differ in different organisms. Further, the distal regions of metacentric chromosomes, specially the subtelomeric regions, are very rich in active genes. Their presence at the subtelomeric region has been confirmed in a large number of plant families, suggesting functionality in relation to specific location below the centromere. This knowledge of the location of gene-rich areas of subtelomeric sites is of great use in genetic manipulation.

Secondary Constriction Region

Nonstaining gaps are observed in certain chromosomes in addition to the primary constriction regions, usually referred to as *secondary* constriction. Since they are usually the sites for the organization of the nucleolus, they are also called *nucleolus organizing regions* (NORs). The region of the chromosome distal to this gap, if a small one, is known as the *satellite* or *trabant*. If a thread is seen to traverse this gap, it is the satellite stalk and the chromosome is a SAT-chromosome. The number of nucleolar organizers in a haploid set of chromosomes varies in different organisms. For example, maize has one and the human cell has five.

The relationship between the secondary constriction region and the nucleolus has been much debated. A large number of workers have suggested that these regions are specifically concerned with the formation of the nucleolus.

Heitz propounded that nucleolus is organized in the destained gap between the satellite and the chromosome arm. McClintock observed in maize that the nucleolus is organized round a deeply staining specialized body – the knob or the nucleolus-organizing element – located at the base of the satellite stalk. Gates, later supported by Pathak, Dearing and others, observed that the secondary constriction regions form small independent nucleoli at telophase, which later fuse into a single large nucleolus. On the other hand, Matsuura, Sato, Darlington, De Robertis and others did not accept the relationship between nucleolus and secondary constrictions, and maintained that all chromosomes, if needed, may form the nucleolus.

The evolution of the *differential staining* methods, including the Feulgen-light green and methyl green-pyronin techniques, which stained the chromosomes magenta and the nucleolus green and vice versa respectively, showed definitely the association of the nucleolus with specific areas, the secondary constrictions of particular chromosomes, from telophase onwards. The maximum number of nucleoli, as observed in telophase of a species, has been found to be constant and to correspond to the number of secondary constrictions present. True diploid species has two identical nucleoli. The number of secondary constrictions and, accordingly, the number of nucleoli may be increased by polyploidy and segmental interchanges. A heteromorphic pair of nucleoli indicates heteromorphicity of the chromosomes.

The secondary constriction region is heterochromatic in nature, though it shows properties different from the other heterochromatic segments of a chromosome. Morphologically, it differs from the primary constriction in the absence of a centromere and in remaining unstained throughout the divisional cycle. When treated with mercuric nitrate or trichloracetic acid and stained subsequently with Feulgen, the primary constriction region is positively stained while the secondary constriction region and the euchromatic segments remain destained (Sharma, 1974).

Evidences for role of NORs in rRNA synthesis

Removal of the NOR may lead to abnormal behavior, often resulting in death. This effect is presumably because the organizer is one of the active sites for the formation of a particular RNA, the ribosomal RNA, without which the lifespan of the cell is limited. The evidences in favor of this concept are the

1. similarity between base ratios of nucleolar and ribosomal RNA,
2. ready hybridization between ribosomal RNA and DNA of the nucleolar region, showing that they are complementary to each other in nucleotide sequence,
3. drastic impairment of protein synthesis in a cell containing a nonnucleolar nucleus, after the supply of ribosomal RNA has been utilized.

Genetic control of nucleolus formation

The genetic control of nucleolus formation has been demonstrated in hybrids of the plant *Crepis*. Each species of *Crepis* has its own specific nucleolus organizer. When two species are hybridized, the organisers show different synthetic ability in the hybrid environment, one being functional and the other remaining inactive by differential repression of genes. When the latter is transferred to the original conditions, it again becomes active.

DNA configuration of nucleolus organizer

DNA in the NOR contains many repeated sequences of DNA (rDNA) which codes for ribosomal RNA; therefore, the region has a high potential for DNA amplification. Electron microscopic studies show the active nucleolar chromatin to have a lampbrush-like configuration. Clusters of RNP fibers project from the DNA and are separated by stretches of naked DNA. RNA–DNA hybridization studies have shown that alternating 18S–28S rDNA sequences are separated by nontranscribed nucleotide sequences, that may correspond to the naked DNA segments.

At the *ultrastructural level,* four main components may be recognized within the nucleolus:

1. *pars granulosa,* containing mainly RNP granules, 150 Å in diameter,
2. *pars fibrosa,* with RNP fibrils, 50 Å in diameter
3. chromatin elements (*pars chromosoma*), and
4. a heterogeneous proteinaceous matrix.

Chromatin elements have been observed in three forms:

1. a nucleolus-associated chromatin, possibly not involved in nucleolus formation, though there are some evidences of its association with certain condensed inactive ribosomal cistrons;
2. septa-like intranucleolar chromatin, often associated with *pars fibrosa;* and
3. isolated or dispersed intranucleolar chromatin threads.

The nucleolar chromatin is a continuous structure and forms a part of the nucleolar chromosome. Nucleoli in normal cells are formed on the same principle as the loops of the lampbrush chromosomes or the functionally active gene loci in interphase chromosomes. A special region of the chromosome, perhaps extending in the lampbrush form or a puff, is responsible for the synthesis of a special type of RNA.

The chromatin responsible for the formation of the nucleolus is its only permanent component. During telophase, the genes located on this segment become active, loops are extended, fibrillar and granular components are synthesized and accumulate in the newly formed nucleolus. During prophase, with the dissolution of the nucleolus, the nucleolar chromatin becomes visible.

The rDNA cistrons, located on the nucleolar chromatin, are mostly redundant and only a few are activated at any particular period. The inactive cistrons are responsible for the condensed appearance of segments of nucleolar chromatin and are late-replicating. The other rDNA cistrons, being in an active and extended state, take part in production of rRNA. The segment where they are located is early replicating. The nucleolus is thus a functional structure and the activity of its permanent structural constituent, the nucleolar chromatin, is facultative in nature. The repetitive ribosomal cistrons have a constant location in the chromosomes. The nucleolar heterochromatin is thus distinct from both the usual facultative and constitutive heterochromatin in that it is a segment with fixed location (constitutive) on particular chromosomes and on it are arranged a cluster of repetitious ribosomal cistrons. These cistrons have an identical, definite but facultative transcriptional property different from that of the unique genome.

Nucleosomes

General morphology

Electron microscopy and biochemical evidences have revealed that the chromosome structure is beaded in appearance and made up of protein octamers, the beads, being termed as nucleosomes (Noll, 1974; Olins and Olins, 1974). There are five histone fractions in the nucleosome assembly, of which four enter into the composition of nucleosome and one is associated with it. Two copies of four histones – H2A, H2B, H3 and H4 form an octamer, with the H1 protein serving as a linker. The octamer is surrounded by a DNA molecule having almost 200 bp, of which approximately 140 bp enter into the coiling of octamers and variable amounts remain associated with the linker H1 molecule. The core of about 140 bp is often highly conserved in nature.

Physical and chemical nature

Nucleosomes are also defined as the core, with the spacer DNA and other proteins associated stoichiometrically with the adjacent core (Rill, 1979). Leaving aside these 140 bp which go into the core, the 40–70 bp of the repeat remain in primary association with lysine-rich histones and nonhistones. From the protein point of view, nucleosome represents the repeats (Sharma, 1984a). In fact, limited nuclease digestion leads to the production of DNA fragments that are integral multiples of 140–200 bp. The relative amount of nonhistone (Hozier, 1979) chromatin is more variable than the amount of histone. It is claimed that in addition to the number of functions that the nonhistone proteins perform in relation to chromosome structure, their role in the organization of chromosome fiber is profound (Hozier, 1979; Sharma and Roy, 1956). The interband DNA in chromatin preparations (Olins and Olins, 1974) consists of a piece of about 30 bp, protected by histone H1, and another piece of 30 bp free of proteins or interacting with nonhistone proteins (Georgiev et al., 1978).

Low-resolution neutron and X-ray scattering studies have shown that the nucleosome is a flat disc rather than a globule. It is about 100 Å in diameter and 50 Å in height. The number of superhelical turns per nucleosome is approximately one/three fourth with a pitch of supercoil diameter of 80 Å unit. The size of DNA in the nucleosomal organization corresponds with the Okazaki fragments of eukaryotes (Kriegstein and Hogness, 1974). Two proteins from the nucleus help in forming the nucleosome assembly, one interacting with the DNA and the other an acidic thermostable protein (originally isolated from Xenopus) that interacts with histones (Lasky and Earnshaw, 1980) (Fig. 1.3).

Function

The nucleosome is considered in the chromosome as one of the first levels of DNA packaging. Initially, the 100 Å thick fiber represents a linear array of nucleosome cores almost in contact with one another. Further coiling of this fiber results in higher order of 200–300 Å thickness. In the 300 Å fibril, the nucleosomes are arranged in a solenoid with 100 Å of pitch

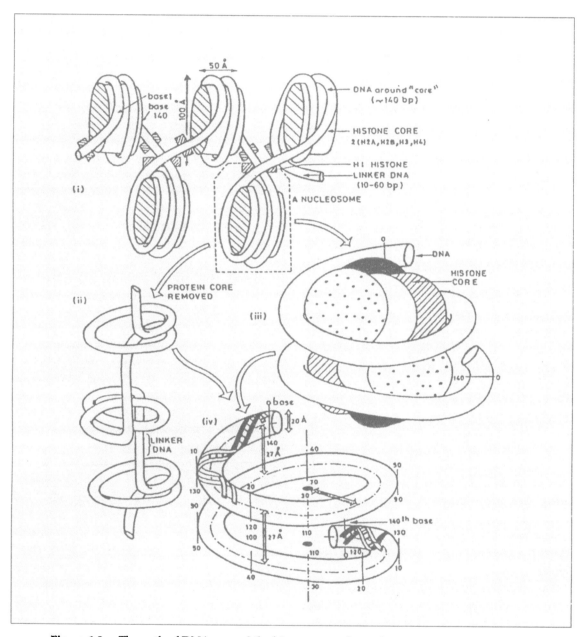

Figure 1.3 The path of DNA around the histone core of a nucleosome.
From Genetics by S. Mitra. Tata McGraw-Hill Publishing Company Ltd. 1994.

(Carpenter et al., 1976). The interaction between nucleosomes is modified or stabilized by H1 histones, which are attached to DNA linker regions between nucleosome cores. In the formation of superbeads, which are aggregates of 6–10 nucleosomes, H1 histones are involved (Renz et al., 1977). In further supercoiling and compaction, the position of the spacer in linker DNA at the outside, accessible to nucleases, is a regular feature.

The structure of the nucleosome reveals remarkable evolutionary stability as noted even in yeasts and humans. Identical molar ratios of four nucleosomal histones and 140 bp of DNA have been recorded. Even the conservation of amino acid sequences in the histones is a part of the stability. The stability of core histones in evolution, along with its highly conserved DNA responsible for structural repetition, is consistent with the apparent inactivity of a vast amount of DNA in the eukaryotic system.

Notwithstanding this conservation, the higher-order chromatin structure and nucleosomal organization are not homogenous along the chromosome (Ruderman et al., 1974). In the plant system, small significant variations have been noted in the nucleosomal organization and length of nuclear DNA between telomere DNA and various repetitive DNA sequences in the chromatin of rye, wheat, and relatives (Venshinin and Heslop-Harrison, 1998). In case of poor fit of linkers between nucleosome as may appear in hybrids, the breakage frequency may increase because of the less efficient repair system in the hybrid background (Heslop-Harrison, 2000).

Nucleosome structure is important for a number of transcriptional phenomena, including (1) silencing at telomere regions, (2) repression of basal promoter activity, and (3) regulation of many inducible genes. Initially it was thought that nucleosome structure acted as a scaffold that was transparent to the transcriptional machinery. It is now clear that multiprotein complexes help the access of regulatory and transcription initiation proteins to the DNA template.

Certain regions of the eukaryotic chromosomes, especially centromeres and telomeres, fail to decondense at interphase. Genes contained in these regions, referred to as heterochromatic, are generally inactive. In yeast, only the silent mating type loci show such a behavior, but lately, it has been recorded that regions adjacent to telomere of yeast (*S. cerevisieae*) have similar heterochromatic properties. The Sir proteins – Sir 1, Sir 2, Sir 3, and Sir 4 – are required for gene expression both on telomeres and silent mating type (Svaren and Horz, 1996) similar to amino termini of histones H3 and H4 of nucleosomes, which are also involved in this type of repression. It has now been demonstrated that Sir 3 and Sir 4 interact with specific silencing domains of H3 and H4 amino terminus. This interaction is also necessary for the perinuclear positioning of telomeric chromatin, a property of heterochromatin. These evidences indicate that regulatory proteins bind to telomeric repeats, which later polymerize into heterochromatic complex and spread throughout the chromosome length interacting with H3 and H4 of nucleosomes. The heterochromatic state, with compact nucleosomes, apparently restricts accessibility of transcription factors to the DNA template.

The factor that influences the structure and accessibility of chromatin is histone H1. Nucleosome positioning is critical in regulating transcription rate by modulating access of the polymerase complex. In general, histone octamers on the DNA show changes in position before and after transcription. Transcription can cause reposition of the histone octamer on DNA or even release of free DNA (Widom, 1997). Association of H1 with core nucleosome particles further inhibits factor binding. It has been shown that histone H1 can inhibit sliding of nucleosomes and, as such, it may play a role in nucleosome positioning. Nucleosomes formed in vitro often occupy different translational frames, but addition of H1 restricts them to certain translational positions. The alteration of H1 as isoforms may influence the specific gene activity during development.

Sir proteins play an important role in silencing but they do not bind to DNA sequences. Recent evidences show that these proteins bind to core histones and histone termini are needed for silencing. These facts clearly demonstrate the role of nucleosomes in gene expression.

Chromosome Scaffold

Before the discovery of chromosomal scaffold by Laemmli and coworkers (1978), Stubblefield and Wray (1971) observed strips of residual structure in isolated Chinese hamster chromosomes treated with NaCl and urea after digestion of the major components of chromosomes, namely DNA and proteins. These authors termed the residual structure as *chromosomal axis* or *chromosomal core*. In the late 1970s, several investigators observed the chromosomal scaffold to be made up of nonhistone proteins (NHPs) in histone-depleted metaphase chromosomes.

There has been a lack of understanding about the roles played by NHPs in structural characteristics of chromosomes. Laemmli and coworkers (1978) put forward the hypothesis of NHP scaffold, based on the electron microscopic observations on surface-spread metaphase chromosomes. It was further demonstrated that the scaffold survived after metaphase chromosomes were digested with DNase prior to the histone depletion. Only a few studies have been done on plant scaffold and most work on non-histone scaffold was carried out in other organisms. Shui et al. (1992) noted the fine structure of mitotic chromosomal scaffold in *Allium sativum* and *A. cepa* under the light and electron microscope, using squash method and silver staining. Electron microscopic observations on the silver-stained structures revealed scaffold as compact fibers and granules, distributed throughout the chromosomes. These observations suggest that the chromosomal DNP (deoxynucleoprotein), consisting of DNA and histones, is combined with NHPs and the latter account for a large part of the combined complex. Cytochemical studies on chromosomes of *A. sativum* indicate that the silver-stained chromosome axis or scaffold is insensitive to DNase, trichloroacetic acid, and sulfuric acid, but sensitive to trypsin and urea treatment. It is thus indicated that the silver-stained scaffold of the plant chromosome is composed principally of NHP. Light, electron microscopic, and cytochemical studies on the kinetochore and chromosome scaffold in *A. sativum* also show that the kinetochore is connected with scaffold, but kinetochore proteins have a higher affinity for silver nitrate than the scaffold proteins do.

Repeated Sequences

Multiple copies of similar DNA sequences were observed in chromosomes in the early 1970s by Crick, Gall, Pardue, and others. The discovery of such multiple copies of similar DNA sequences in chromosomes is a major event in the study of chromosome research. Following DNA denaturation and subsequent reannealing at lower temperature, repeated sequences show strong reassociation kinetics as compared with lesser repeats or unique sequences. These repeats may be highly homogeneous, as in satellite DNA sequences, noted initially in *Xenopus* by Pardue and Gall, or may be moderate or minor in nature. Such sequences may also be *inverted* in nature, as in *palindromes* (Cavalier Smith, 1976). In the chromosome structure, highly homogeneous repeats may be located in one locus, whereas minor or moderate repeats may be interspersed throughout the intercalary or terminal positions (Sharma, 1985). The demonstration of the repeated sequences accounts, to a great extent, for the huge amount of DNA noted in the different organisms. The significance of such DNA sequences can be judged from the very fact that in several plant species, especially in grasses including barley (Ranjekar et al., 1974), 72–90% of DNA is repetitive in nature (Sharma, 1984a). This higher repetitive content is often responsible for the *C-value paradox*, as exemplified by the presence of approx. 20–100 pg of DNA in lilies and amphibians, as compared to 0.2 pg in *Arabidopsis* and approx. 7 pg in man (Scherwood and Patton, 1982).

Between plants and animals, there are some fundamental differences in genomic composition vis-a-vis chromosome structure. In animals, dispersed repeated sequences in chromosomes are region-specific, which is not so in plants. In mammalian species, each chromosome shows a characteristic GC–AT ratio, whereas chromosomes in widely different plant genera, viz. wheat, tomato, and field bean, show more or less similar nucleotide composition. In general, repeat DNA in these two major groups behaves differently. Plants show greater similarities between all chromosomes in a species.

Types of repetitive DNA

The length of repetitive sequences may vary from simple sequence repeats of dinucleotides through tri-, tetra- and hexanucleotides to nucleosome repeats of 180 bp and upto 10,000 or more base pairs in rRNA sequence families. The distribution of these sequences is chromosome-specific, species-specific, and sometimes variations are there at the intraspecific level, indicating their role as parameters of genetic diversity. These sequences generally fall into two categories: (1) tandem repeats, each sequence arranged tandem to the other forming a monomeric unit and (2) other group–highly dispersed along with other sequences, scattered throughout the genome. They include mobile elements such as the long interspersed nucleotide element (LINE), the short interspersed nucleotide element (SINE), the long terminal repeat (LTR), and other dispersed repeats. On the basis of repeats in different organisms, certain clear distinctions have been noted in their occurrence in plants and animals.

In the plant system, as the repeats show high homogeneity between all chromosomes in a species, unlike in animals, the study of marker sequences becomes more difficult in plants than in animals.

In a large-scale chromosome organization in plants, functional genes are organized in clusters mainly in subtelomeric sites. Plant chromosomes show characteristic linear distribution of dispersed repeat, but homogeneity has been noted with respect to specific genome, unlike in animals. Plants, in general, have a wide distribution of simple sequence repeats (SSRs). Such SSRs flanked by DNA sequences, present once in genome, are termed *microsatellites*. Repeat sequences occur throughout the entire length of the chromosome, but their presence in specific locations such as pericentromeric, subtelomeric, telomeric, and intercalary regions is universal. They may be simple sequence repeats, minisatellites, satellite, and in ribosomal DNA. Some of the simple sequence repeats present once in a genome, flanked by DNA sequences, as mentioned above, can serve as good markers. The highly homogeneous repeats of the satellite DNA are mostly AT-rich.

Certain tandemly repeated sequences may be genus-specific or unique to a particular species. Their distribution and copy number help in differentiating various linkage groups in the chromosome complement. Two different repetitive sequences covering almost the length of one nucleosome are present in tetraploid *Brassica napus*. Such sequences are present in diploid ancestral species namely *B. campestris* and *B. oleracea* as well. These sequences can also be used as markers.

Location

Highly homogeneous repeats have been located in the ribosomal RNA genes. These genes in chromosome serve as useful markers, especially the highly conserved 5S,18S, to 5.8S, and 25S rRNA gene loci. In Triticeae, 18S–25S and 5.8S sequences show different rearrangements and

deletions in barley, rye, wheat, and allies, and provide clues to genomic rearrangement. Their polymorphism in different accessions of cowpea has been noted, suggesting their value even as genotype diversity at intraspecific level.

The major sites for highly repeated DNA sequences have been identified by in situ hybridization in the telomeric regions in rye and in the pericentromeric regions in barley chromosomes. This location conforms with the distribution of heterochromatin in the corresponding chromosome regions. Knob-heterochromatin of *Zea mays* too is rich in satellite DNA content. The major component is a 180-bp repeat (Dennis and Peacock, 1984). The same sequence is present in *Tripsacum dactyloides*, but this species differs from maize in other repeat sequences where the complexity of this satellite DNA is in the range of 200 bp. Telomeric heterochromatin too is often rich in repeated sequences. In the NORs involving 80S and 20S rDNA, highly repeated sequences have been recorded; repeats are also well known in the 5S sequence.

A characteristic repeat (GGGGATTT) is found at the end-points of chromosomes. Such sequences in different plant species are largely localized mainly at the telomeres, near certain centromeres as in *Arabidopsis thaliana* or at major intercalary sites in the median region of chromosome arms. In *Allium*, it is a complex array of a 175-bp repeat.

Location of highly-repeated sequences has been confirmed in pericentromeric regions, which show condensed heterochromatin and low density of active genes.

Accessory chromosomes in a number of plant species such as *Scilla autumnalis*,

Secale cereale, S. montana, and *Brachycome dichromosomatica,* which are the cytological embodiment of heterochromatin, are rich in highly repeatcd sequences. Both moderate and minor repeats more or less remain interspersed in between the unique sequences.

Evidence has been obtained for the interspersion of short repeats (mainly 200–400 bp long) in a large portion of the cereal genome. The proportions of the genomes with this kind of organization have been estimated as 40–50% for oats, 50–60% for hexaploid wheat, and 30–35% for rye. Most of the genome of these cereal species is occupied by palindromes and tandemly arrayed repeats. These are of two types: (1) very similar repeats and (2) complex combinations of different repeats.

Over and above such interspersion, the presence of repeated copies in intervening sequences or introns, universally in transposons, as well as in flanking regions of replicons has been noted.

Evolution

The mechanisms suggested so far for the origin of repeated sequences include saltatory replication, unequal crossing over, and transposition including insertion. Accumulation of short repeats is a very common event in the cycle of evolution and it may even represent duplication of regulatory sequences. A certain part of the genome displays marked variability when compared to its other parts. The unique sequences, at least those in the regions concerned with the encoding of metabolic enzymes, have stable structures possibly maintained by natural selection. The repeated sequences, on the other hand, are most susceptible to change resulting from amplification, translocation, deletion,

mutation, and other tools of evolution, leading to novel genomic configurations. Sequences in new environments may undergo patterning and amplification, leading to new repeat families.

Several repeated sequences, both coding and noncoding, often show intraspecific homogeneity. This behavior, termed as *concerted evolution* of repeated sequences, suggests that related repeat families in different species have diagnostic sequence variation (vide Brown, 1984). During evolution, a progressive homogenization of repeated sequence families through continued turnover of sequences has been suggested. Such a process can be mediated by gene conversion, transposition, and unequal exchange. Notwithstanding the continued divergence of different plant groups, repeat sequences of each have undergone a concerted evolution. It is quite likely that different mechanisms have been involved in origin and evolution of other repeated sequences, which might have withstood the rigors of selection.

Significance

Ample evidences have been accumulated to show that evolution has been associated with the increase or decrease in the DNA content (Price, 1976) in which amplified sequences play an important role. Amplification can bring about an increase in the total DNA content in the cell as well as in small chromosome regions, thus disproportionately increasing the size of the regions. Evidences indicate that increase in the amount of DNA in the cell with concomitant increase in its size may lessen the rate of mitosis and meiosis (Bennett, 1982). High DNA concentration in a chromosome segment may have an undesirable outcome.

Several nonspecific functions involving cell and nuclear size, cell and nuclear volume, chromosome cycle, generation time, duration of meiosis, and the padding to keep the chromatin in the folding state have often been attributed to repeated sequences, which have otherwise been termed as *nucleotypic DNA* (Bennett, 1972). The role of interspersed repeats in repair synthesis has been demonstrated in *Lilium* (Davidson and Britten, 1979), where the function is performed by moderate repeats during pachytene by conserving such DNA sequences (Lavania and Sharma, 1984). The importance of repeats in the regulation of chromosome structure in folding during pairing is on record. In replication, as worked out, a subset of dispersed repeats may act as initiation points in early development and differentiation (Taylor, 1984). In gene clusters with flanking repeats, repeats may be involved in code expression through methylation of specific genes during a particular phase. One of the important evidences of the regulatory role of the repeats is derived from fission yeast. In maize, repeated sequences are associated with specific classes of heterochromatin having particular meiotic effects. Their conversion, both in nucleotide sequence and chromosomal location, suggests a selective function.

Repeated sequences have also been visualized as reservoir of new sequences and loci of accumulation for mutation (Sharma, 1983). Their interspersion may possibly allow flexibility of units to a certain extent. The moderately repeated classes of DNA sequences coding for proteins are located in the genes coding for histones. The presence of multiple copies may meet the requirements for rapid histone synthesis during embryogenesis (Davidson, 1976). A possible role of repeats in DNA replication, especially at chromosome ends, has been suggested by various authors (Harfman et al., 1979; Heumann, 1976).

Direct evidence of involvement of repeats in gene conversion and compression has been gathered in fission yeast (Egle, 1981). Intergenic conversion and reappearance of suppression alleles are facilitated by exchanges in repeats. In *Saccharomyces purpuratus* (Lewin, 1980), RNA hybridization is always preferential, where nonrepeats are associated with repeats.

The role of palindromic repeats is diverse. They are replicated under specific conditions. Protein synthesis is affected at different levels through such sequences. Functions attributed to them include recognition systems, both at DNA and RNA levels, involving deletion and translocation, cleavage sites, termination of transcription, binding of regulatory proteins, and attachment of chromosomes with each other helping the information transfer (Lavania and Sharma, 1984). In fact, the presence of palindromes in transposons in gradients and increase in higher organisms suggest a distinct functional role.

The occurrence of repeated sequences in introns – the intervening sequences – in genes, facilitates alternate splicing or reshuffling of exons and intergenic conversion. In transposons, such sequences permit dispersions and promote genetic diversity through mobility. Repeats have been extensively observed in spacer regions in several organisms. In vivo experiments indicate their role in initiation and termination of transcription as well as their association with heavy RNA synthesis.

Role in molecular documentation and DNA fingerprinting

DNA fingerprinting has immense potential specially in documentation and analysis of biodiversity. It is based on the fact that each individual has a unique DNA pattern. Polymorphism in genome can be detected by variable number of repeat sequences arrayed in tandem sequence. In addition to determination of genetic diversity, molecular hybridization through synthetic oligonucleotide probes, complementary to short tandem repeats, can trace the pathways and trends in evolution.

In general, in plants, *short tandem repeats* are common in the genome of which bi- and tetranucleotides are most abundant. However, their sequence, location, and frequency often give a clear picture of the trend in evolution. For example, the species-specific nature of STRs is indicated by one in every 11 kb in *Petunia*, one in every 25 kb in *Brassica* whereas it is one in every 42, 58, and 156 kb in *Arabidopsis*, *Zea*, and *Hordeum* respectively. In rice, there is a variety of STRs whereas only AT type STR is found in *Hordeum*. In dicots, STRs are three times more than in monocots. In algae, all STRs contain GC base pairs, whereas it is 50% in monocots and 15% in dicots. The nature of sequence is distinct in assemblages; their frequencies in broader taxonomic groups and differences in different plant groups indicate their role as landmarks in evolution and phylogenetic status. Their occurrence at specific locations is not random, but follows a distinct genetically controlled pattern. Their clear significance in evolution is yet to be established.

Accessory chromosomes

Large amounts of repeated DNA sequences have been recorded in *accessory* chromosomes, which are often regarded as cytological embodiment of repeated sequences. The behavior of accessories on the spindle is rather erratic. The correlation of such accessories, associated with adaptation in different environmental set-up, has been recorded in different plant species (Müntzing, 1977). Their effects on cross-over are also known. The adaptibility of accessories in certain sub alpine species, such as *Arisaema*, as well as experimental demonstration of their adaptability in *Allium stracheyii* are also on record (Sharma, 1979). Moreover, these chromosomes have the capacity of independent replication, irrespective of other chromosomes, as noted in pollen grains of *Anthoxanthum*, *Poa*, and other genera.

Accessory chromosomes, as the evidences go, are mostly additional genetic elements without structural genes. In that event, to some extent, they are comparable to self-replicating plasmids of prokaryotes despite their location inside the nucleus. There are ample records of B-chromosomes or accessories thriving outside the nucleus forming micronuclei. Their rapid multiplication has been demonstrated in several plant species. As in the case of plasmids, with judicious manipulation in suitable materials, B-chromosomes may serve as vectors for transferring genes from one organism to another. Such a possibility is indicated by the fact that they may sometimes carry euchromatic or more precisely, structural genes.

Repetitive DNAs in chromosomes are additional gene sequences which, due to their property of amplification, dispersion, and mobility, confer flexibility to the species. They do not necessarily involve structural genes and are present in multiple copies. As such, their manipulation at such sites may not lead to deleterious effects on the organism. Their function as spacers, noncoding elements, or insertion sequences, makes them ideal sites for restriction enzyme operation. Several of the repeated segments are involved in intra- or intergenic conversions. Different restriction sites too have been located within a single repeated segment. These sequences are ideal sites for genetic manipulation in chromosomes.

Split Gene

The complexity in the uninemic, multi-repliconic and beaded structure of the chromosome is further reflected in the demonstration of split nature of the gene. The single gene responsible for a polypeptide is split in nature and contains informative and noninformative sequences, termed as *exons* and *introns* respectively, lying adjacent to each other. The best evidence of the presence of such sequences was obtained from the fact that the primary transcript is much longer than a translatable transcript. Posttranscriptional events include scissoring of introns and joining of exons by restriction enzymes and ligases. This splicing property is very characteristic of the eukaryotic system. It adds a new dimension to the gene–enzyme relationship because through alternate splicing, a variety of mRNA transcripts may originate from a single gene sequence.

The identification of split genes has been further confirmed through various techniques, including isolation of intact DNA sequences of a gene, cloning of complementary DNA, preparation of probes, restriction analysis, ultrastructural analysis, and sequencing DNA. A large number of genes have been reported in eukaryotes, of which a major fraction belongs to the mammalian system. In mammals, *Neurospora*, and yeasts, introns have highly repeated DNA, mostly of GT–AC sequences. Introns or such intervening sequences may occur in multiple copies at intercalary, terminal, or initiation points. In repeated sequences, such dispersed introns may be related, varying in length, or may not be related.

As regards exons, plant and animal chromosomes do not differ to a great extent; 140-bp sequences are the most frequent ones, except 200 exons of globin genes. A remarkable correlation is often noted between the exon sizes with 50, 140, and 200 bp and the length of the DNA associated with nucleosome linker, core particle, and an entire nucleosome respectively.

Sizes of exons may vary widely from 0.01 kb to 4.9 kb in certain cases, though smaller introns of 100–200 bp are quite frequent. In general, genes smaller than 0.5 kbp do not have any introns, the best example being histone H4 genes. As such, the size of introns is controlled by the genetic information content of the exons. Introns have been regarded as insertion sequences as well, as recorded for *jumping or mobile genes*. It is claimed that recombination within introns may help in the rearrange-ment of exons, as in hen lysozymes. Their differential processing may assist in the reshuffling of exons.

The antiquity of introns, also supposed to have originated from transposons, has been demonstrated in different groups of plants and animals. Introns of active genes in maize and soybean are more or less located at the same site, as also in fungi, protozoa, and animals. While comparing the arrangement of mouse and rabbit, it has been shown that the length and position of the intervening sequences between the two segments of DNA represented in the messenger, are almost the same. The major introns are 550–600 nucleotides long in both cases and occur adjacent to the sequences coding for 104 amino acids in mouse and between 100 and 120 in rabbit. It indicates that the process that led to the formation of the split gene occurred at least 5×10^7 to 5×10^8 years ago, and the DNA has remained stable.

All these evidences clearly indicate that introns cannot be considered as recent insertion sequences, but rather they have had a long antecedent period of evolution and might have evolved at a very early stage of eukaryote.

Similarly in the study of phylogeny and plant evolution, the significance of intron sequences is paramount in relation to providing clues to ancestry. In tracing the ancestry of land plants, the homology of the intron sequences of Coleochaete and *Chara* with those of *Marchantia* has been regarded as the signal evidence, indicating also the conserved nature of intron sequences through geological ages.

Mobile Sequences

The discovery of mobile sequences or transposons has added a new dimension to chromosome structure. Mobile elements are specific DNA sequences having the capacity of movement from one location to another in the chromosome. The movement of transposable elements or the mobile DNA sequences throughout the genome is effected through a process of excision and reintegration (Chandlee, 1990). Such sequences were initially located in maize by McClintock (1956), which has a high degree of repeats comparable to jumping genes. Later, these have been found in several organisms, including yeast, which may cover a major fraction of the plant genome.

Primary Structure

Mobile genetic elements of maize have the ability to transpose autonomously or under the influence of autonomous elements present elsewhere in the genome. A regulator or receptor is as a rule, incorporated in the MGE (mobile genetic element) system, which may also consist of a set of different regulators and receptor elements (vide Shumny and Vershinin, 1989).

The maize activator (Ac) – Dissociator (Ds) system seems to be the one most studied in molecular terms. A brief characterization of the primary structure of the Ac–Ds system is as follows:

The Ac and Ds elements have inverted sequences at their ends and they generate duplications at their insertion site. Two Ac elements independently isolated from two different *waxy* mutants show a length of 4563 bp (Muller Neuman et al., 1984; Pholman et al., 1984). The Ds elements differ from the Ac elements in a central deletion, which may abolish the ability for autonomous transposition. The Ds elements have been isolated from the *ux*, *sh*, and *Adh 1* mutants and deletion lengths vary from 194 bp to 4.531 bp (Doring et al., 1984). The termini of Ds consist of 11 nucleotides TAGGGATGAAA; Ac has the same primary structure as Ds with the difference that the outermost nucleotide A is replaced by C in Ac. A salient feature of these two elements is the presence of many direct and several inverted repeats of different lengths. It is considered that the organization of the repeats in the Ds and Ac is not random, though its significance is unclear. The Ac can be double and a copy may be inserted in inverted orientation. The phenotype can revert to normal, completely or partially, after removal of the MGE from the locus, into which it has been inserted.

The MGEs have been discovered and investigated in other plant species, besides maize like. *Glycine max* (the Tgm element); *Antirrhinum majus* (the Tam 1 and Tam 2 elements). These species are taxonomically (i.e. evolutionarily) quite distant and they belong to different families. They show remarkable similarity in their structural organization (Venetski, 1986). Thus the elements Spm (maize), Tgm 1, and Tam 1 have almost identical terminal inverted repeats; Spm contains the hexanucleotide ACACTC and Tam 1 contains the septanucleotide ACATCGG within its structure. All these three elements produce a duplication involving three nucleotides in the target gene. The Ds-like sequences in the genomes of rye, wheat, and barley vary from 100 to 300, the range being less than that of the maize genome. Differences in the variation range among *Trillium*, *Secale*, and *Hordeum* are rather small.

There are certain structural features common to the MGEs of animal and plant cells. Homologous terminal sequences have been observed in the maize MGE Cin 1 and in the *Drosophila copia* elements (Shepherd et al., 1984). Evidence of this structural parallelism indicates its nonrandom nature.

Transposability

Maize MGEs undergo *transposition* from one chromosome to another (Finchman and Sastry, 1974). However, the following are certain characteristic behaviors of transposition elements of the maize (Khesin, 1985): (1) The greater the distance between the sites of transposition, the more rare is their transposition occurrence; (2) the once transposed elements, after moving, can return again to the original locus, thereby generating a second mutation: and (3) the same element can transpose to distinctly different genetic loci. For this reason, there is no doubt that a plant containing a mobile element can despatch it to any other gene locus. Transposition is very site-specific.

In the case of precise excision elements, the genes concerned can revert to their original state, and the stable and wild phenotypes are restored. If the excision is imprecise, the locus remains altered but becomes stable. The adjoining segments of the chromosome strongly affect transposition behavior of the elements like any other event of genetic recombination (Khesin, 1985). Transposition frequency is also related to the distance between MGEs and heterochromatin. The proximity of heterochromatin has a profound effect on MGE transposition and stabilization.

Types of mobile elements

Transposable sequences have been classified into two distinct categories: (1) insertion sequences that are short, about 1000 bases and (2) longer transposons, which may even be several thousand bases long.

The distinction of dispersed mobile repeats is also based on their mode of amplification and movement through DNA or RNA intermediates termed *transposons* and *retroposons* or *retrotransposons*, respectively.

In general, transposons are associated with the gene coding for transposition – *transposase* – and a sequence recognized by transposase. In certain cases, only the recognition sequences are present.

Retrotransposons, in contrast, are endowed with code for their own reverse transcriptase, RNA binding capacity, and integration into the genome. These retroelements, which are highly dispersed with other sequences, fall into two categories, depending on their size and location: they may be long terminal repeats (LTRs) or long interspersed nucleotide elements (LINEs). LTR transposons are also called viral retrotransposons because of their close similarity to them. There are other retroelements that encode only transcriptase activity. Simultaneously, some of the dispersed repeats, short interspersed nucleotide element (SINEs), are retroelements that originate from reverse transcriptase activity, elsewhere in the genome.

In plants, though several transposons have been recorded (Smith, 1991), majority of them fall into two categories: (1) in Ac/Ds of maize and Tam 8 of *Antirrhinum majus* and (2) En/Spm of maize and Tam 1 of *Antirrhinum*, Tgm 1 of soybean, and Pis I of pea. But none of these transposons amplify to more than a few hundred copies in the genome. The size too is small, the largest being 15 kb in the Tam1 transposon.

In yeast (*Saccharomyces cerevisiae*), 35 dispersed copies of the repeat family, such as the Tyl gene, have been recorded, constituting nearly 2% of the genome. In culture, there has been a distinct shift in location. The movement of transposable elements in yeast is spontaneous and does not require activation by mutagens. The Tyl of yeast has mostly a repetition of 258 base pairs with 100 copies of the gene.

Function

Transposition is associated with marked chromosome rearrangements. Activation of transposable elements may generate stable or unstable single gene mutations, methylation changes in and eventual activation of transposable elements, as well as additional chromosome breakage (Peschke and Phillips, 1992). Ac and Ds elements produce mutations at the *shrunken* locus encoding endosperm sucrose synthase, at the *waxy* locus encoding glucosyl transferase, and also at other distinct loci. Insertion of these elements has minor effects on the viability of the mutants. Autonomous transposable element activation may be a symptom of genetic instability, as in tissue-culture-generated plants or their progeny. Such elements move from one position to another, interrupting gene function when inserted into genes (Peterson, 1993). Insertion of transposable elements in a host site adds nucleotides to the genome and induces a target site duplication.

Altered nucleotide sequences result from an excision process involving a protein–DNA complex in which endonuclease enzyme excises the element affecting the gene reading frame and altering the template sufficiently to change the coding of protein (Peterson, 1993).

Mobile repeats because of dispersion contribute to a great extent to the total DNA content of the genome. The size of the genome associated with adaptation is, to a great extent, dependent on the functioning of dispersed repeats. Several dispersed repeats show mobility and distribution indicating retroposon activity. As they have their own integrase and reverse transcriptase, they are not dependent on other factors for survival. The possibility of interspecies or even intergroup transfer is indicated in the similarity in mobile sequences of Ty1 of yeast and copia in other organisms as well. The mechanism of horizontal transmission is not fully known but the transmission through viral capsid has been regarded as a possibility (Smith, 1991). Retroelements with repeats widely dispersed in the chromosome constitute a major fraction, and their amplification along with transposons contributes to variation in genome size.

The action of transposable elements has been shown to be controlled by specific *proteins* having the capacity to recognize the ends of the elements (Vide Sharma, 1985). When these proteins are provided, the ends are activated and different target sites are invaded, leading to transposition and other DNA arrangements. One protein found to be necessary for transposition has been termed "transposase."

The transposition proteins evidently bind to the ends of the elements as also to other proteins, thereby inhibiting cleavage of DNA chains at the point of junction of element and the host DNA. Demonstration of the presence of an active protein and its control over transposition have opened up the possibility of using them in artificial gene transfer with predictable accuracy. In gene transfer, or site-directed mutagenesis, transposon mediation is one of the most powerful tools.

Significance in evolution

Transposition of mobile genetic elements has a direct bearing in the evolution of species in biological systems. Their mobility, pleiotropic effects on surrounding genes, presence of hot spots for integration, and capacity to transform other sites into cohesive hot sites are properties that suggest evolutionary progress (vide Sharma, 1985).

It has been shown that recessive character in pea in Mendel's experiment was in fact an insertion in the normal dominant gene. It has been noted that wrinkled character, which is recessive to dominant round, is due to the absence of branched starch and amylopectin. Normally, the round seeded variety has the enzyme SBE (starch branched enzyme) responsible for branched starch and formation of amylopectin. This enzyme is absent in the wrinkled seeded variety. Further molecular studies have shown that the gene coding for enzyme is about 3.3 kb long in the dominant variety, whereas in the recessive (i.e., in the wrinkled seed), sequences are present and have a length of 4.1 kb. This difference in length is mainly due to a long jumping sequence of 0.8 kb being inserted within this gene, thus making it nonfunctional. This is a very good example of a jumping sequence making a gene nonfunctional, and its apparent recessive nature is due to this insertion. This fact further shows repeated sequences and especially jumping sequences have played an important role in evolution.

Evidences have also been gathered indicating the importance of transposon in adaptation. It has been suggested (vide Moffat, 2000) that larger genome with high DNA acquired through amplification of retroposons may help plants to meet stress situation by enabling plants to seek or retain water.

Isochores: Gene Space

In chromosomes, genomes are now regarded to be made of *isochores*, which are very long stretches of DNA with compositional homogeneity. Genes are embedded in DNA fragments in chromo-somes, which differ 1–2% from each other in GC content. In humans, genes are present in a few GC-rich fractions corresponding to 3–4% of the genome. Almost all genes in maize are present in isochores with a narrow GC range (1–2%) covering 10–20% of the genome. The hypothesis of "gene space" has been put forward, which implies genome regions represented by a single family of isochores. Gene space in chromosomes of maize is the only genomic compartment where mobile sequences can be transposed. Foreign DNA transferred within these GC-rich regions of chromosomes is likely to be transcribed, whereas insertions in other regions may fail to do so. In view of GC-rich regions of isochore in chromosomes, it is suggested that AT-rich sequences in transgenes may be the markers for identification and integration in chromosomes (Kamptla et al., 1998). Existence of isochores and their function suggest that in the chromosome, a genomic system exists to identify and inactivate sequences that are compositionally not matching with the chromosome segments serving as integration sites.

CHAPTER 2

Chromosomes and Cell Division

Mitosis and Meiosis

The behavior of chromosomes follows a distinct pattern in the body, termed *mitosis*, quite different from that in germ line, the *meiosis*. In mitosis, chromosomes undergo longitudinal division, each half going to one daughter cell. Following reconstitution at the end, the same number is maintained throughout the somatic cells. At prophase, chromosomes appear longitudinally split along the entire length. Early prophase shows the uncoiled nature of the chromosome thread. With the progress of prophase, chromosomes start coiling or undergo spiralization leading to condensation, which reaches a high level in metaphase. With the initiation of metaphase, along with chromosome condensation which is maximum, nuclear membrane disappears and spindle is formed with two poles. Chromosomes arrange themselves at the equator, being attached to the spindle at the centromeric region. Gradually, with the onset of anaphase, the two longitudinal halves of the chromosomes, earlier held together by the centromere, move towards the two poles with the aid of the microtubules. The centromere helps the attachment of the chromosomes with the spindle and their movement towards the poles. The movement is facilitated through the motor protein *dynein* at the centromeric locus. Finally, at telophase, the two daughter halves of cell are separated and groups of chromosomes at the two poles organize themselves inside the nucleus, within the nuclear membrane. The result is the formation of two daughter nuclei. These daughter nuclei are ready for metabolic phase. Nuclear division is followed by the division of cytoplasm or *cytokinesis*, leading to formation of two daughter cells. Equational separation in normal mitosis is responsible for every cell containing the same genetic component. This type of division ensures equal number of chromosomes in all cells of the soma or body.

Another type of division characterizing the germinal line is *meiosis*, where the chromosome number undergoes reduction. This division occurs in male and female organs of flowers as well as in testis and ovarian systems of animals. Meiosis entails two divisions of the nucleus, one of which is equational and the other is reductional. It characteristically results in the separation of homologous chromosomes and halving of the chromosome number.

Of the two divisions in *meiosis*, in the first division, prophase is divided into five phases: *leptotene, zygotene, pachytene, diplotene*, and *diakinesis*. In leptotene, chromosomes are long, thin, and optically single threads, showing little of the coiling characteristic of somatic prophase. The threads contain a series of chromatic beads called *chromo-meres*. In zygotene, the homologous chromosomes begin to pair, usually at the ends, near the kinetochore or both. A general pairing is followed by closer chromomere to chromomere association or synapsis which may be complete or not, depending on the species and on condi-tions within the organisms. During pachytene, interchange of segments takes place between the paternal and maternal sets of chromosomes through a process known as *synapsis* or crossing-over. The chiasmata, which is the visible sign of crossing-over, involves only two out of four strands at any locus. Breakage and reunion of segments are achieved during this phase, and the two chromosomes originating thereby contain interchanged segments (Fig. 2.1). Of the two chromatids present in each chromosome, one repre-sents the original gene component and the other recombinant one. At the diplotene, contraction and the opening out between the homologues continues, and chromo-

somes tend to clump in the center of the cell. Formation of a major coil usually begins at this stage.

At diakinesis, contraction is near the maximum and the chromosome pairs are well spread throughout the cell by mutual repulsion. The pairs of homologous chromosomes are still held together, but there are a number of cases of continued association in which no chiasmata can be demonstrated. The nucleolus generally disappears during this stage, but may persist. During the first division of the metaphase stage, the two homologous chromosomes form a bivalent lying on the equatorial plate, with homologous kineto-chores oriented toward opposite poles. At *anaphase*, homologous kinetochores move towards opposite poles. In *telophase*, there is regrouping of chromosomes at the poles. In the next stage, the *interphase* (i.e. interkinesis stage between two divisions), a cell wall is laid down between the two nuclei to give a two-celled dyad structure (photographs 2.1–2.9).

In the second division, prophase is essentially a stage of contraction and coiling. In metaphase, the kinetochores, which have been holding the two chromatids together, line up at the somatic metaphase and divide. In anaphase, sister kineto-chores separate to the poles, pulling with them the chromatids to which they are attached. In telophase, there is reconstitu-

Figure 2.1   Diagram of crossing over as it takes place in meiotic prophase through the breakage and reunion of non-sister chromatids at the four-strand stage.
From Cytogenetics by C.P. Swanson.

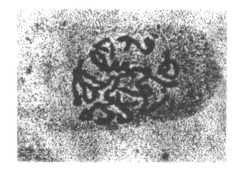

Photo 2.1 Prophase

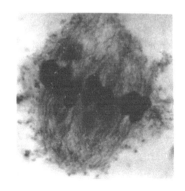

Photo 2.4 Metaphase (side view)

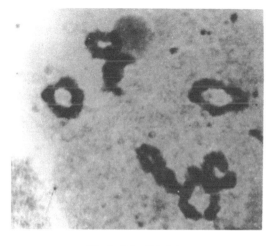

Photo 2.2 Diplotene

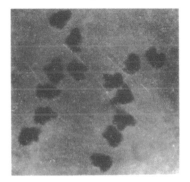

Photo 2.5 Anaphase (early)

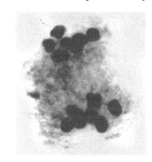

Photo 2.6 Anaphase (late)

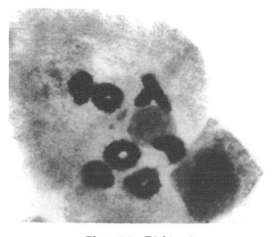

Photo 2.3 Diakinesis

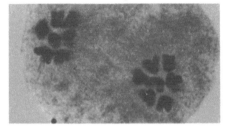

Photo 2.7 Metaphase II (polar view)

Photo 2.1 – 2.9 divisional stages in meiosis. **2.2 – 2.7** *Hordeum vulgare* (n = 7) **2.8, 2.9** *Dipcadi serotinum* (n = 8)

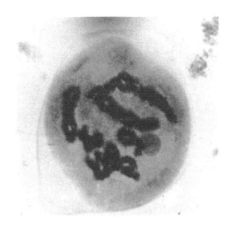

Photo 2.8   Diplotene

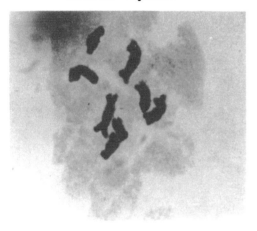

Photo 2.9   Metaphase (early)

tion of interphase nuclei and the laying down of cell walls to give four cells, known as *tetrads*.

Synaptonemal complex

The synaptonemal complex (SC) is considered as the ultrastructural manifestation of meiotic chromosome pairing and synapsis. It is found in meiotic prophase between two paired homologues represented in 10-nm fibers arranged in superstructure. They are considered to be necessary for synapsis to align the two homologues, pairing at the molecular level. It is very prominent in the pachytene stage. There are lateral elements and the central element connected by the transverse elements, which connect the central core with lateral elements. It has the appearance of a ladder, and each synaptomer is supposed to be spaced at minimum of 20–30 nm.

In lateral elements, both the nucleic acids and proteins have been found, but the central core is rich in RNA, in addition to the much lower amounts of DNA and protein. The SC as such is a tripartite body located between the synapsed chromosomes. There are two lateral dense fibers twisted around each other, with a flanking central ribbon. The lateral arms are joined to the adjacent chromosomes by fibrils about 10 nm in length. The microfibrillar material arises from the chromosomes themselves, whereas the synaptonemal complex is proteinaceous, consisting mainly of scaffold proteins. The SC is present in the pachytene from one end to the other of the bivalent (Gilles, 1981, 1983). With the aid of the spreading technique, at higher magnification both lateral and central elements can be studied. Chromatin attached with SC is highly dispersed and normally a background network becomes visible. In maize, the length of SCs had been found to be variable, and along with arm ratio, the bivalents can be identified.

With the transition from pachytene to diplotene, the SC becomes very short. The lateral elements become convoluted and the central elements appear as dots or become absent. Often the lateral elements form branches and forks. It is often possible to locate nodules in the central region of some SCs, which might indicate recombination nodules.

In several organisms, it has been claimed that the presence of perfectly paired homologous chromosomes at pachytene does not reflect perfect synapsis of homologues during initial pairing and synaptonemal formation at zygotene. However, in case of wheat, it has been shown that majority of zygotene pairing is between homologous chromosomes and chromosome segments.

Chromosome architecture and the process of condensation

Early attempts to elucidate the process of condensation suggested that there are major changes in the phosphorylation of histones as cells enter mitosis. A kinase that performs phosphorylation, the complex of Cdc2 and cyclin B, initiates principal biochemical activity inducing mitosis. Despite this correlation, however, the importance of histone phosphorylation in mitotic chromosome condensation remained unclear. Using *Xenopus* egg extract as the experimental system that lacks transcription activity, it has now been shown that a five-subunit protein complex formed as "condensin" is essential for mitotic chromosome condensation. Condensin introduces positive supercoils into DNA in the presence of topoisomerase I and adenosine triphosphate. Comparison of the condensin complex from interphase and mitotic extracts reveals that three of its subunits of the condensin complex become phosphorylated in the mitotic extract and that only the mitotic form of the complex has the ability to supercoil DNA. Cdc2 is likely to be the kinase that phosphorylates and activates condensin. This may trigger mitotic chromosome condensation. Its depletion would lead to decondensation.

However, substantial regions of chromosomes that constitute heterochromatin remain condensed throughout the division cycle, including interphase.

Evidence of crossing-over: interchange of segments and recombination

Every meiotic division as suggested above consists of two division cycles – reductional (known as heterotypic division) and equational (known as homotypic division). After reductional telophase, the dyads enter into homotypic division and each daughter nucleus contains half the number of chromosomes. Prior to the onset of prophase of this homotypic division, there is no further division of chromosomes. As such, during the succeeding metaphase of meiosis, i.e. metaphase II, each chromosome, composed of two chromatids, one pure and the other recombinant, arranges itself at the equator and the typical mitotic separation follows. Thus, the tetrad originating out of dyads results in four spores, containing haploid or half the number of chromosomes. Of the four nuclei, two are recombinants. These spores ultimately give rise to gametes containing only one set of chromosomes or haploid ones. When the male and female gametes, each being haploid, unite, the zygote is formed with diploid set of chromosomes. Throughout the organization of the body, equational division is the general rule followed by reductional separation during the formation of germ cells.

Meiotic division is the principal factor for incorporating recombination in the offspring. It is remarkable that in the regular pattern of equational and reductional separation which apparently leads to monotony in the life cycle, the scope of generating variability through recombination is inherent in the system.

Thus, during meiosis, two events are of crucial importance: 1. synapsis: the interchange of segments and 2. reduction in the number of chromosomes to form the gametes.

As mentioned in the preceding paragraphs, two chromosomes of the bivalent remain intertwined and each point of twining is termed as chiasma. Chiasma is the visible sign of crossing-over. The issue of chiasma as an index of crossing-over was initially debated and views were expressed that crossing-over takes place after the formation of chiasma. The issue has long been resolved that crossing over precedes chiasma formation. At meiosis, the evidence of interchange of segments occurring during crossing over was demonstrated by Creighton and McClintock (1931) in maize as well as by Stern in *Drosophila*. The basic principle in their experiments was to secure a heteromorphic pair in which the two chromosome ends of one member are distinctly different from the other. Such chromosome combination was obtained by Creighton and McClintock in securing a maize hybrid in which two chromosomes of homologue type 9 were different from each other. One of them had a knob at one end and a part of chromosome 8 translocated at the other end, making it distinctly longer than its homologue. The other member was thus comparatively short and knobless (Fig. 2.2). Such heterozygote had colored aleurone and starchy endosperm. The genes were located in two chromosomes, both being dominant over the recessive colorless and

waxy endosperm character. On crossing the heterozygote with double recessive (i.e. colored aleurone and waxy endosperm), four types of progeny were obtained. Two were of the parental type and the other two were new combinations (i.e. colored nonwaxy and colorless waxy endosperm). The new combinations also showed the presence of two new chromosomes (i.e. *knobless long* and *knobbed short*) which could have arisen only by crossing-over.

Variations in Mitosis

Variations in the mitotic process may be induced by a variety of chemical and physical stimuli. Multiplication of chromosome strands without mitosis appears to be rather common, especially in nuclei of differentiated cells. Geitler (1953) published an extensive review of the subject of multiple chromosome replication (polyteny). In 1947, Huskins provided experimental evidence of the existence of such strand multiplication in plant roots by showing that the average number of chromocentres was correlated with the degree of polyteny. Occasionally, such cells enter division either spontaneously or by induction but, instead of going through a series of "reduction" divisions, the nuclei become typically polyploid, with each chromosome having the usual two chromatid strands. Another kind of "somatic reduction" division was reported by Huskins (1948) following treatment of plant meristems with sodium nucleate. He noted that at stages from prophase to metaphase, there was a strong tendency for whole chromosomes to separate into two groups. This behavior, followed by more or less normal anaphase separation of chromatids, leads to formation of a multinucleate cell and with further

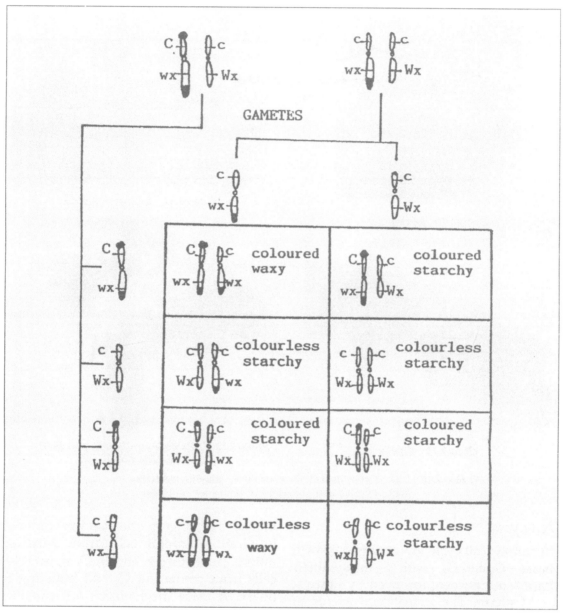

Figure 2.2 Creighton and McClintock's demonstration of exchange of segments between homologous chromosomes with cytological markers which help to identify the recombined chromatids. The cytologically observed exchange was found to be associated with crossing over of genes at the two loci.

cytokinesis, and ultimately to cells with less than diploid number of chromosomes (photographs 2.10–2.13). However, similar reduction figures have also been obtained (Bhattacharya and Sharma, 1952) in untreated onion roots.

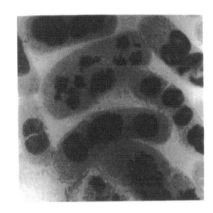

Photo 2.10   Multipolar

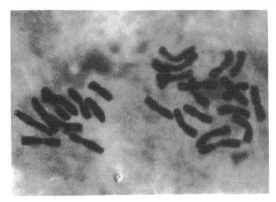

Photo 2.12   Bipolar - unequal

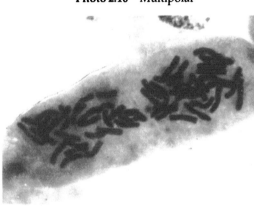

Photo 2.11   Bipolar

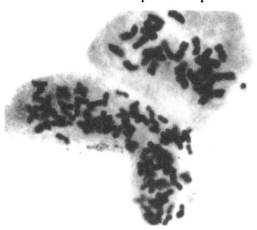

Photo 2.13   Cells showing diploid/polyploid
chromosome numbers

Photos 2.10 – 2.12   Groupings of chromosomes - somatic reduction
2.13   Diploid and polyploid chromosome numbers

Cell cycle

The mitotic cell cycle involves two growth phases – G_1 and G_2 – with S – the synthetic phase – in between, followed by mitosis, the M phase. The transitional events of DNA replication and mitosis are promoted via the activity of complexes of cyclins and cyclin-dependent kinase (cdks). There is an interdependence of these transitions, which involves two types of mechanisms. One generates a pathway of events acting on G_1, which are required for S phase, and events acting on G_2, which are required for

mitosis. The other mechanism uses controls in G_1/S to inhibit cells from entering mitosis prematurely, and in G_2 to prevent cells from reentering G_1 and initiating S phase. If these mechanisms fail and the interdependence of these events is lost, the result would be a change in ploidy level. Successive divisions without an intervening S phase normally result in reduction in ploidy level and nonviability. But such a situation is normal for meiosis where reduction in number is desired. Similarly, successive S phases without intervening mitosis results in ploidy increase, such as endoreplication and endopolyploidy common in endosperm suspensor cells (Connell and Nurse, 1994).

Check-points permit monitoring of progression through the succession of essential steps and thus coordinate the cell cycle for replication and segregation of components.

One of the important discoveries in recent years has been the cyclin-dependent kinase, which is necessary for cell cycle. The enzyme kinase, which controls the behavior of other proteins by attaching phosphate groups, and the protein cyclin, which triggers the kinase at a critical period bringing about changes in cell cycle, are both essential. Cells engaged in the mitotic cell cycle passing through the G_1 phase are required to promote the onset of S phase and prevent the onset of mitosis.

Two important events are clearly involved in eukaryotic gene replication: (1) at the end of mitosis and (2) at the beginning of S phase. The initiation competence at G_1 is capable of triggering "S" phase. The initiation competence is lost in the "S" phase with the onset of replication. Further reversal of competence starts from G_2 to G_1 through the M phase. The competence evidently is established much before the "S" phase, though not adequate to initiate replication. At the G_1/S boundary, the process reaches its culmination, triggering S phase and DNA replication. With replication, there is again loss of competence.

Cells use three mechanisms to ensure the accurate transmission of their genetic information: (1) repair mechanisms that correct spontaneous or environmentally induced errors in DNA replication and chromosome alignment; (2) delay mechanism that detects errors and arrests the cell cycle until repairs are complete; and (3) induction of the death of damaged cells as a way of preventing them from giving rise to mutant progeny. The term "cell cycle check-point" refers to the entire process of monitoring cell cycle events, such as DNA replication and spindle assembly, generating signals in response to errors in these processes, and halting the cell cycle at a specific point (Murray, 1991).

The *start* is defined as the point in G_1 after which cells are committed in mitotic division. The major cell cycle transitions are controlled by cyclin–cdk complexes. Experiments in fission yeast show that cyclin B is required not only for a cell to traverse G_2 and enter into mitosis but also for a cell to perceive and act on the G_2 phase. Thus, the presence of cyclin B–cdc2 complexes is essential. In absence of such complexes, cells are redirected into G_1 and undergo an additional S phase (Fisher and Nurse, 1996).

In yeasts and *Aspergillus*, a single cdk is required to traverse the G_1/S and G_2/M boundaries, with specificity being provided by the formation of complexes with different cyclin molecules. Higher eukaryotes use different but related cdk subunits for controlling these cell cycle transitions (Forsburg and Nurse, 1991; Vide Sharma, 1999).

Spindle

The spindle structure, associated with chromosome movement, shows in polarized light to be fibrillar in nature with at least two types of fibers: (1) the traction or chromosomal fiber, extending from the centromere to the closest pole and (2) the spindle or continuous fibers, running parallel to each other from pole to pole. They constitute the body of the spindle. A third type of fiber has been visualized, arising between the chromosomes immediately after separation in anaphase – the interchromosomal fibers. Their existence is, however, debated.

Two forces are mainly involved in the poleward movement of the chromosomes after they divide (1) *a pushing force*, in which the central region between the split chromosomes, after they divide elongates, pushing the chromosomes to the opposite poles, and (2) *a pulling force*, which draws the chromosomes to opposite poles, caused by the shortening of the chromosomal fibers.

The chemical nature of the spindle has been studied to a certain extent and reveals the presence of protein, nucleic acid, carbohydrate, lipid, ATPase, free sulfydryl groups and zinc. Electron microscopic studies have also shown several components within the spindle, including a certain number of microtubules associated with chromosomes (Salisbury, 1995).

The microtubule cytoskeleton, of which the spindle is a part, is involved in generating and maintaining cell polarity. Many different morphological structures can be generated by only a few structural elements. Microtubules are basically hollow cylinders made of tubulin and various associated proteins. Microtubule associated proteins (MAPs) from different organisms and tissues have been extensively studied.

Spindle fibers are regarded as long chain fibrous proteins, rich in sulfur, extending from pole to pole. They are composed apparently of small protein molecules associated end to end by S–S (disulphide) linkages. The globular protein of the cytoplasm, through possibly the action of the spindle substance, is converted into fibrous protein of the spindle. Experimental evidence of this structure was obtained from experimental treatments with compounds which break the S–S linkage, like mono- or dimercapto-propanol or digitonin. Following this treatment the spindle is not formed. Colchicine was observed to exert a similar effect. Attempts were made to isolate the spindle from sea urchin eggs in which mitosis was blocked at metaphase with colchicine. The matter thus isolated was an amorphous gel without visible fibers. It was suggested that spindle formation took place in two stages: (1) formation of a gel through polymerization by intermolecular S–S bonding: this stage is not affected by colchicine and (2) the formation of secondary bonding and orientation of spindle protein into the fibrous system: colchicine inhibits the secondary bonding. The ATPase activity appears to be associated with the fibrous component of the spindle.

The relation of the spindle to alignment of chromosomes in metaphase and their movement in anaphase has not yet been explained satisfactorily. It is possible that the spindle fibers are actually attached to the centromere to provide an oriented field for forces inherent in the chromosomes, thus contributing to their movement. Electron microscopy has shown chromosomal fibers to develop from the centromere and to grow towards the corresponding pole. Similarly the loss of acentric fragments during anaphasic separation, indicates an actual attachment of chromosomes to the spindle.

CHAPTER 3

Karyotype Concept

In the genome, differences in chromosome morphology between different genotypes also indicate differences in the genic contents of the individuals in general. The major variations that can be observed by comparing related species can be divided into (1) variations in chromosome morphology; (2) variations in absolute chromosome size; (3) variations in relative chromosome size; (4) variations in chromosome number including aneuploidy, euploidy, and heteroploidy, and (5) variations in staining properties.

Variations in size are influenced by either differences in amount of gene products or proteins produced by the individuals, or duplications of genes, which influence their interactions and the rate of synthesis of individual proteins. Variations in staining properties at the intra- or interchromosome level are usually indices of differences in the timing of coiling or replication cycle. Changes in chromosome morphology may involve alterations in gene arrangement, which may influence their subsequent segregation and recombination. Alterations in chromosome number may lead to differences in gene content and arrangement or gene duplication or deficiency or both. Chromosomal differences thus reflect basic differences in the source or genome while morphological, physiological, and biochemical differences indicate variations in the products of gene action as also modified by environmental factors.

The Russian school of cytologists, headed by S. Navashin, developed the fundamentals of the *Karyotype Concept* from their observations that most species of living organisms show a distinct and constant individuality of their somatic chromosomes and that closely related species have more similar chromosomes than those of more distantly related ones. The karyotype was first defined in 1926 by Delaunay as a group of species resembling each other in the morphology and number of their chromosomes. Since the Liliaceous genera on which he worked such as *Ornithogalum* are rather homogeneous, the term appeared to coincide with the conception of a genus. However, Levitsky, on the basis of refined methods, indicating that the evolution of karyotype in many genera takes place through a series of alterations in chromosome morphology, gave a new definition for karyotype. According to him, karyotype is the phenotypic appearance of the somatic chromosomes in contrast to their genotype. Amongst plants, their appearance is usually recorded from metaphase stages in the somatic meristems. In certain groups such as Bryophytes and also in some materials like Liliales, karyotypes can be studied from gametophytes such as the pollen grain.

In comparing karyotypes of different species, certain fixed criteria are usually taken into account: (1) variation in absolute chromosome size; (2) variation in the position of the centromere; (3) variation in relative chromosome size; (4) variation in basic number; (5) variation in number of chromosomes with secondary constrictions and positions of satellites; and (6) variation in the degree and distribution of heterochromatic regions and repeated DNA segments (photographs 3.1–3.9).

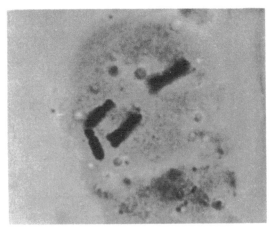

Photo 3.1   *Haplopappus gracilis (2n=4)*

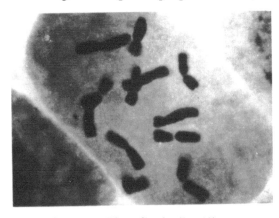

Photo 3.3   *Rhoeo discolor (2n=12)*

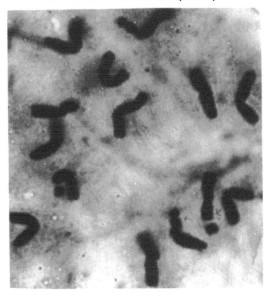

Photo 3.4   *Hordeum vulgare (2n=14)*

Photo 3.1 – 3.8   Root tip cells showing karyotype of different species.
3.9   Karyotype of *Nothoscordum fragrans* from pollen grain cell (2n = 22, n = 11).

Photo 3.5 *Allium cepa* (2n=16)

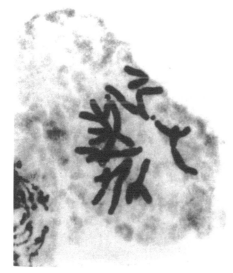

Photo 3.6 *Allium fistulosum* (2n=16)

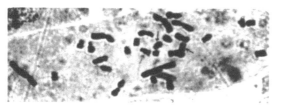

Photo 3.7 *Scilla indica* (2n=30—bimodal karyotype)

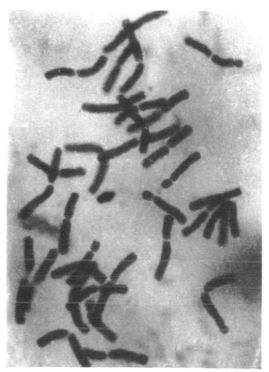

Photo 3.8 *Allium tuberosum* (2n=4x=32—autotetraploid)

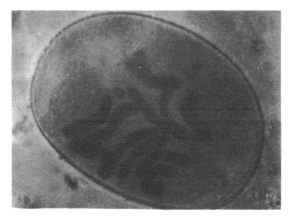

Photo 3.9 *Nothoscordum fragrans:*– division in pollen grain (n = 11)

Evolution from taxon to taxon in general has always involved one or more of these characteristics. Their primitive or advanced nature has been judged on the basis of their occurrence in species, the taxonomic positions, and relative antiquity, which are well established through other parameters. All these changes, except that involving absolute size, reflect changes in the chromosome structure and number.

The diagrammatic representation of the karyotype is termed as an *idiogram*. Idiograms are very useful for comparison of karyotypes of different species or for the study of evolutionary trends.

Trends of Karyotype Evolution in the Plant Kingdom

Variations in absolute chromosome size

Variations in absolute chromosome size are apparent between different groups as a whole, but also, in some cases, between different members of the same group. The total mass of chromosomes in a nucleus has been found to be closely related to its DNA content. At any particular stage of development, the ratio of DNA to proteins is relatively constant in eukaryotes and indicates the amount of genic material present.

The distribution of these values usually follows certain regular patterns. The smaller taxonomic units such as species or genus generally show a similarity in DNA content within a range, however, with exceptions. Within each broad group, there is much variation. Sometimes the range of variation within classes, orders, or even families may be far greater than that between classes. A factor responsible may be polyploidy, repeat DNA content, or increase in the basic number. Nevertheless, often genera within the same family show great differences in chromosome size, examples being the two members of the family Leguminosae, *Glycine max* and *Vicia faba*, the DNA content of the latter being nearly six times that of the former. Despite such variations between smaller units of classification, mean and modal values for DNA content in different groups of vascular plants are rather similar.

Regularities

Certain regularities recorded include

1. Heterosporous taxa, amongst the spore-bearing vascular plants such as *Selaginella* and members of Marsileaceae, tend to have smaller chromosomes than the homosporous ones.
2. Amongst the major groups of plants, gymnosperms usually have the largest mean and modal size of chromosomes. Certain angiosperms, such as *Paeonia, Lilium, Tradescantia* and the mistletoe genus *Phoradendron* may have even larger chromosomes.
3. Chromosomes of woody angiosperms are usually small with little size difference between related species and genera. Exceptions include the family Annonaceae.
4. Chromosomes of herbaceous angiosperms show great size difference between different genera of the same family, sometimes even between different species of the same genus, e.g. *Oxalis*.

Hypotheses

Several hypotheses were proposed earlier to explain these differences in size:

1. A higher amount of DNA in the nucleus is regarded to indicate a higher number of significant genes, leading to greater diversity of gene function. According to this hypothesis, evolution from simple to complex forms has taken place by the

addition of functional genes. Evidences from homologous gene sequences in related proteins show that this hypothesis may explain satisfactorily the evolutionary origin of new enzymatic function. This theory, however, cannot explain wide differences in DNA content observed between species of the same genus, as in *Oxalis*, or the chromosome volume per nucleus being higher in the more primitive seed plants, such as the Cycads, than in the majority of the highly advanced angiosperms.

2. Species with larger chromosomes possess larger quantities of DNA with "nonsense" sequences of nucleotides, having no adaptive value. A consequence of this view, would be that in the older taxa, species with an initially large amount of DNA, would gradually lose it through random deletions. Thus, individuals within the same species and related species, separated geographically, would show variable chromosome size. However, this effect is not seen in actual cases. For example, all diploid species of *Trillium* — a genus existing for 40 to 50 million years — have five pairs of very large chromosomes. Although the species are scattered over the temperate areas of the world, yet their chromosomes resemble each other in gross morphology and size. It is obvious, therefore, that the large amount of DNA present in these chromosomes is not useless or junk, but has some adaptive significance.

3. Species with large quantities of DNA have multistranded chromosomes and are therefore genetically equivalent to polyploids, though the somatic number may be diploid. Comparative measurements of DNA in Droseraceae and Ranunculaceae have been taken to indicate that in certain cases polynemy may be responsible for increase in chromosome size during evolution. However, these suggestions are yet to be confirmed with additional experimental data.

4. Presence of large chromosomes may indicate the occurrence of gene loci that are replicated many times in series, tandem fashion. Such duplications are regarded as adaptive. Rees and his group, from studies on *Lolium*, *Lathyrus*, and *Allium*, suggested that increase in chromosome size is due, at least in part, to the lengthwise replication of chromosome segments. From a comparison of breeding systems and morphology of phenotypes, chromosome size has been found to increase in *Lolium* and *Allium* and to decrease in *Lathyrus*. Such replication of chromosome segments has been regarded to provide opportunities for the differentiation of genes with new function and the establishment of lateral heterozygosity.

The tandem duplication hypothesis does not refute the multiple-strand hypothesis, but rather it complements the latter. Both forms of duplication may have been effective in increasing chromosome size during evolution. Duplication of gene sequences can account for the reverse trend to decrease in size as well. In certain highly specialized forms, decrease in chromosome size occasionally takes place, as for example in species of *Crepis* and *a Muscari*. This trend can be explained in *Crepis* (Compositae) by assuming that the evolution of annual growth cycles reduces the adaptive value of extra duplications of certain gene loci, resulting in their elimination. Similarly, in *Muscari* (Liliaceae), the evolution of reduced inflorescences may be associated with smaller chromosomes, occurring from the loss of gene loci, being rendered useless.

Frequency

Phylogenetic reduction and phylogenetic increase in chromosome size have been found to be about equally common in higher plants and the processes are reversible. In certain groups, there is no correlation between geographical distribution and chromosome size. In others, however, species with large chromosomes occupy cooler climates than those with smaller ones.

Variations in basic number

Alterations in the basic series may or may not be accompanied by polyploidy. In higher organisms, they are usually the end product of a series of changes in the x or basic number, followed by different levels of polyploidy. Aneuploids are more frequent than direct euploids or amphidiploids.

The most common alterations in chromosome number in higher plants are duplication or higher multiplication of the entire chromosome complement, often with structural changes. In several plant genera, species form a polyploid series. For example, in *Dioscorea*, the species are multiples of $x = 10$, from $2n = 20, 40$ to 120. Usually lower numbers are more primitive than higher ones as seen from taxonomic characteristics. This trend is usually irreversible. However, occasional reduction from higher to lower ploids may also take place. *Euploids* are polyploids that are entirely multiples of a single basic number.

In many cases, multiple series of chromosome numbers are present in species that have arisen by doubling of chromosomes from hybrids between diploid species having different basic numbers leading to amphidiploidy. For example, amongst the cultivated species of *Brassica*, *B. napus* shows $2n = 38$ in its body cells. The number has arisen by the duplication of the entire chromosome complement of a hybrid between *B. campestris* ($2n = 20$) and *B. oleracea* ($2n = 18$).

Aneuploid variations form a series in which the gametic numbers of related species are present as a consecutive series, or differ from each other by two or more chromosomes. For example, the species of *Crepis* form a series of $x = 3, 4, 5, 6,$ and 7 chromosomes.

Evolution may involve a descending series as in *Crepis*, where $x = 7$ is present in the most primitive species and successively advanced ones have progressively lower basic numbers. Such alterations usually occur through successive unequal translocations, which also progressively increase the differences in relative size between the chromosomes. In cases where the smallest chromosomes have heterochromatic regions near the centromere, where the genes show relative inertness, all the essential genetic material can be removed from such chromosomes through a final unequal translocation. As such,

such a truncated chromosome, with only a centromere and adjacent heterochromatic regions, can be easily lost, leading to a reduction in the basic number of chromosomes. This method for decrease in basic number through this mechanism is referred to as *dislocation*.

Increase in basic chromosome number has more elaborate mechanisms. Chromosomal rearrangements involved can be studied from organisms subjected to radiation or other chromosome breaking agents, where structural heterozygosity may arise through reciprocal translocations.

A possible method is by *nondisjunction of a small bivalent* during first meiotic metaphase, so that one gamete contains both chromosomes. An offspring of this gamete with a normal one would be a plant trisomic for the small chromosomes. Usually such individuals are unstable. However, they may survive in certain cases, particularly in interracial or interspecific crosses, for indefinite periods. Reciprocal translocation may occur between such a small chromosome and a larger one, so that essential genetic material is transferred to the former. In such cases, the new chromosome becomes an essential component of the karyotype, leading to an increase in the number.

Occasionally, in artificially induced translocations, a structural heterozygote may have a small fragment with a centromere and attached chromosomal materials derived from one of the normal chromosomes. If another translocation leads to transfer of some more essential material to such a chromosome, it persists, causing an addition to the number in the basic set.

Another type of aneuploidy similar to a progressive increase in chromosome number may arise by a progressive decrease or increase in the basic number followed or accompanied by *amphidiploidy*. For example, in *Crepis* spp. with $x = 7, 6$, and 5 chromosomes, different polyploid and amphidiploid combinations can produce every haploid number, from $x = 10$ upwards. Polyploids of $x = 5$ can give rise to $x = 10$; $x = 11$ can arise from amphidiploids of hybrids between $x = 5$ and 6; $x = 12$ from $x = 6 + 6$ or $7 + 5$; $x = 13$ from $7 + 6$, and so on (Fig. 3.1). A very extensive aneuploid series is the one observed in the genus *Carex*, with haploid numbers from $n = 6$ to $n = 56$, which have possibly arisen by structural changes of chromosomes followed by autoployploidy or amphidiploidy. However, this genus, and others of the family Cyperaceae, are characterized by the presence of diffuse centromere, where the property of spindle attachment is distributed throughout the length of the chromosome. All fragments produced in such chromosome complements have a chance of survival, giving rise to various numbers. Translocations, hybridization between species with different numbers, and subsequent chromosome doubling can result in the extensive aneuploid and heteroploid series observed in these genera.

Thus, naturally occurring aneuploid series of chromosome numbers in sexually reproducing plants can be classified into (1) descending basic, e.g. *Crepis*; (2) ascending basic, e.g. *Allium*; (3) interchange polyploid amphidiploid, e.g. *Brassica, Stipa, Carex*; and (4) unbalanced numbers after polyploidy and subsequent apomixis.

In the short-term evolution of plants, both auto- and allopolyploidy may have been dominant factors, since polyploidy increases the "biochemical versatility" at

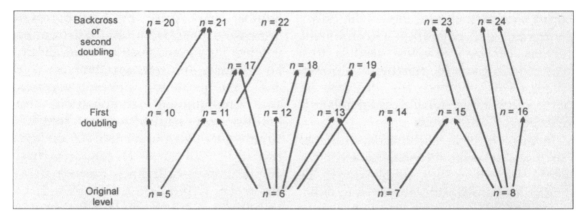

Figure 3.1   Various methods through which successive hybridization and chromosome doublings can lead to different aneuploid chromosome numbers (after Stebbins GL 1971; *Chromosome Evolution in Higher Plants*. Edward Arnold, London).·

the primary enzyme level. It can create new "hybrid" enzymes through oligomerization of related polypeptides produced by related (homologous) genes. Such enzymes, with different properties, may extend the range of environments in which normal and successful development may occur.

Cryptopolyploidy refers to an increase in chromosome size, as shown by a series of doublings of minimum genome size, indicated by DNA/genome ratio, rather than an increase in ploidy level.

Variations in form and relative size of chromosomes

These include variation in chromosome segments and heterochromatin.

Symmetry and asymmetry

Variations in the size and shape of chromosomes have been observed, to a greater or a lesser degree, between species of the same genus and between genera belonging to the same family, having the same chromosome number. In a large number of cases, such changes show positive trends, which can be correlated with trends observed in phenotypic differences. Based on such observations in the members of the tribe Helleboreae of the family Ranunculaceae, Levitsky suggested that evolution in the form and shape of chromosomes has resulted in progressive asymmetry of the karyotype. The most primitive genus in Ranunculaceae, judged on the basis of floral characters, was found to have a karyotype with large chromosomes of more or less the same size and with median centromeres. Such a complement was regarded as a *symmetrical* karyotype. In the most advanced genera of the group, *Aconitum* and *Delphinium*, the more advanced species show progressive asymmetry in the karyotype with variation in size. It was suggested that two trends can be observed in the progressive evolution from a *symmetrical* karyotype to an *asymmetrical* one: (1) reduction in length of one arm of the chromosome, giving rise to submedian and subterminal centromeres from median ones; and (2) reduction in size of some chromosomes in relation to others of the same set, so that the advanced karyotype has chromosomes of progressively unequal sizes.

Thus, a *symmetrical karyotype* will have chromosomes of more or less similar size, with the chromosome arms of almost equal length. An *asymmetrical karyotype* has, on the other hand, many chromosomes with subterminal centromeres or greater size difference between different chromosomes or both.

Karyotypes of the different species of a genus may either be alike or may vary greatly between species. Even in those genera where the karyotypes of different species are very similar, such as *Paeonia* and *Lilium*, meiotic studies of F_1 interspecific hybrids indicate that they differ from each other in inversions and translocations. These differences could not be identified in the karyotypes through conventional methods because the inversions are paracentric ones and the translocations involve small segments.

In some species, morphological changes may involve alterations in the number of chromosomes with secondary constrictions, as in *Lilium* with progressive polyploidy. Alterations in the patterns of distribution of heterochromatin have been found to accompany speciation in North American species of *Trillium*.

Trends in evolution

Most of the herbaceous genera with medium-size to large chromosomes show variations in karyotype between species. Usually, variations may follow any of the four following patterns:

1. Increase in asymmetry accompanied with decrease in chromosome number and increase in specialization of certain morphological characters, e.g. *Crepis* spp;

2. Increase in asymmetry accompanied by an increase in chromosome number and certain specialized characters, e.g. *Clarkia* and *Brodiaea*. This pattern is much less frequent than the first one;

3. Occurrence of different basic chromosome numbers with little or no change in the symmetry of the karyotype, e.g. species of *Ranunculus* and *Allium*; or

4. Constancy in basic number accompanied by variation between species in karyotype symmetry, e.g. species of *Aegilops*.

There is a predominant trend in flowering plants towards increasing asymmetry of the karyotype, usually associated with greater specialization in different aspects. This trend can be observed clearly on comparing the more advanced genera such as *Aconitum* and *Delphinium* having highly specialized flowers with the more primitive ones such as *Nigella* in the tribe Helleboreae of the family Ranunculaceae.

Morphological effects of structural changes on the karyotype depend on the location of breaks and reunions and the genes involved. Since symmetrical changes need more precise location, obviously asymmetrical changes are more frequent, leading to progressive asymmetry of the group.

The process of *centric fission*, or the breakup of a metacentric chromosome into two telocentric chromosomes, assists in progressive asymmetry. However, the opposite process, that of *centric fusion* between two telocentrics or reciprocal translocation between two acrocentrics, may correspondingly result in increasing symmetry of individual chromosomes. The former process would give rise to a metacentric and the latter to two metacentrics, one large and one small. Survival of both the new chromosomes would result in increasing asymmetry amongst the members of the karyotype, giving rise to *bimodality*. If the small product is lost, as in species of *Crepis*, the basic number is reduced as also the bimodality. If further evolution takes place through subsequent polyploidy and structural alterations, it becomes very difficult to trace the evolution and interrelationships of such taxa. For example, in the Cycadaceae and Podocarpaceae, where differing frequencies of stable telocentrics are found, opinions are divided as to whether evolution has been through centric fission or vice versa. The process from asymmetry to symmetry due to centric fusion has been recorded in the genus *Gibasis* of the family Commelinaceae.

A method to identify secondary trends towards decrease in asymmetry from primary trend of increase would be a study of the process involved. Increased asymmetry usually results from pericentric inversions and unequal translocations of chromosome segments without changing the number of centromeres or chromosomes. The change involved is that of the fundamental number of the large chromosome arms. Decreasing asymmetry by centric fusion, on the other hand, leads to a reduction in the number of chromosomes though the fundamental number of chromosome arms remains unchanged.

Adaptive significance

Two hypotheses have been advanced to explain the adaptive significance of karyotype asymmetry during evolution:

1. The *ease of separation theory* is based on cell mechanics. It suggests that the small or the telocentric chromosomes are more likely to separate earlier than the large metacentrics, during prometaphase and metaphase, and thereby have a greater chance of survival. Also, in the small chromosomes, heterochromatic regions flanking the centromere are smaller and thus capable of separating more quickly than those of larger chromosomes.

   The fallacies of this hypothesis lie in the fact that since prometaphase and metaphase are relatively short phases, they are unlikely to affect the entire divisional cycle. In addition, if this concept is accepted, there should be a tendency to eliminate all large metacentrics. However, highly evolved genera such as *Lilium* and *Delphinium* show two large metacentrics in their karyotypes. Thus, the concept appears to be untenable in view of the existing data.

2. The *linked gene cluster hypothesis* visualizes the presence of linked clusters of genes on the longer arms of chromosomes within an asymmetrical karyotype. Beneficial genes continue to be added to them by inversion and translocation, whereas the arms without such clusters would continue to lose genes through deletion and become progressively shorter.

In invading new habitat, a species can achieve dominance, either by acquiring asymmetrical karyotype, cross-fertilization to grow new forms, or by self-fertilization to hold together the clustered genes, followed by polyploidy which confers adaptability. Simultaneously, apomixis helps to retain the desirable characters at the initial period.

The extent to which chromosome characteristics independently can indicate the direction of evolution of a group has been questioned by Keith Jones. He has advocated a synthetic approach, viewing the chromosomes in conjunction with all other characters of the organisms in determining the pattern of evolution. Stebbins, however, suggested that the question be resolved on the basis of the relative frequency of certain phenomena observed during the process, trends towards asymmetry being more common than their reversal.

Apomixis and its role in evolution

In the stabilization of karyotype in evolution, the role of apomixis needs to be discussed. Apomixis, by definition, is propagation without undergoing mixing for fertilization. It is well known that this type of reproduction is very common in ferns, where a haploid gametophyte can give rise to a haploid sporophyte (apogamous sporophyte) without fertilization and a diploid sporophyte can give rise to a diploid gametophyte without reduction division (aposporous gametophyte).

In case of phanerogams, there are examples where fertilization and reduction divisions are avoided and the diploid nature of sporophyte (plant itself) is generally maintained through unreduced gamete, i.e. without reduction division. Formation of diploid nucleus without reduction division can also occur through restitution of nucleus without daughter nuclei formation.

Apomixis can be *vegetative* or *sexual*. Vegetative apomixis implies reproduction through cuttings, bulbs, corms, bulbils, rhizomes, or leaf, as in *Bryophyllum* and others. Vegetative reproduction, if associated with chromosomal changes or mutation, may give rise to new genotype and mutants as in species of Liliaceae, Araceae, and other monocots. Sexual apomicts may arise from diploid tissues of sex organs such as anther wall, ovular cell, nucellus, and other such tissues. Polyembryony from such tissues is very common as in species of *Citrus*. The unreduced gamete from the megaspore mother cell too may lead to direct seed formation without meiosis as in species of *Taraxacum* (Koltunow, 1993). Similarly, fusion of two synergids, i.e. two haploid nuclei in the embryo sac mother cell, without fertilization, also comes under sexual apomicts. Even haploid plant may arise from the egg cell without fertilization, leading to haploid apomicts.

All such methods of propagation come under the category of apomixis. Such apomicts are noted in nature and can be produced artificially for genetic improvement of crop species. Apomixis may be *obligate* or *facultative*. Obligate apomicts can only reproduce through asexual means and do not have the capacity for sexual reproduction. Facultative apomicts can resort to sexual reproduction when needed as in species of *Potentilla* in Rosaceae and Gramineae. This category of species is very successful in evolution as they have retained the capacity of generating variability through sexual reproduction. One of the genetic reasons of most of the species of Gramineae to become so successful in evolution is their capacity of facultative apomixis. The grasses, with more than 10,000 species, are successful because of various reasons: high frequency of hybridization, high frequency of polyploidy giving resistance to stress and adaptation capacity to different conditions, and large number of cytotypes with numerical and structural variations, along with facultative apomixis. Genetically, various processes of apomictic cycle are controlled by different genes and a delicate genetic balance among a series of factors leads to successful apomictic reproduction (Stebbins, 1950).

In nature, apomixis plays an important role in stabilization of polyploids and hybrids. Polyploid species, on their origin, often show irregularity in gamete formation and aberrant meiosis. This results in sterility, and often extinction of the genotype. As such, natural polyploids often resort to apomictic type of reproduction, i.e. without meiosis, till normal meiotic behavior is restored through gradual mutation of the gene, controlling pairing, and multivalent formation. Control of multivalents, normal bivalent

formation, and regular meiotic behavior lead to adaptation and survival of plants. That is the reason why natural auto-tetraploids with multivalents are very infrequent except in *Allium tuberosum*, *Galax aphyllum*, and others. Similar apomictic behavior also helps hybrids to tide over the period of aberrant meiosis and gradually become stable in evolution.

The question is often raised whether apomixis is a *progressive* or a *regressive* force. It is beneficial as long as environmental conditions remain constant. But under a changed environmental setup, apomicts not having the capacity of generating variability through meiosis are often ineffective in with standing rigors of selection. As such, it is a regressive force in the long run, though effective for a short period, in the evolutionary timescale. Partial apomicts, which have the capacity of resorting to sexuality are, therefore, considered to be most successful in evolution. If needed, they can undergo meiosis and lead to recombinants under stress situation or a changed setup. In the origin of new and variant karyotype and its stabilization in evolution, the role of apomixis is paramount.

CHAPTER 4

Chromosomal Changes and Biodiversity

PART I

The study of biological diversity is essential for plant scientists. In the plant kingdom, the spectrum of diversity is very wide occurring at every level of taxonomic hierarchy. The most convenient analysis of biodiversity lies at the level of phenotypes. Phenotypes may visibly differ in morphological, anatomical, as well as physiological characteristics. Environmental factors play an important role in the origin and stabilization of diversity. However, diversity in the true sense is a reflection of the individual genetic characteristics. Excepting monozygotic embryos, no two individuals are alike and as such, diversity can be traced back to the molecular nature of genomes and even a single nucleotide difference can bring about diversity. In practice, the study of diversity at present ranges from the phenotypic features to genetic characteristics at an intraspecific level. Biodiversity, which, in the true sense, is genetic diversity, is measured at the level of chromosomes in the genome. The most common mechanism of origin of diversity is gene mutation. In addition to point

mutation, genetic changes can be brought about through alterations at the chromosome level, both *numerical* and *structural*, which are the basic mechanisms of origin of diversity. Over and above these alterations, change in the position of gene at an intrachromosomal level by different processes, can cause diversity, which is otherwise termed as *position effect*.

Numerical Changes

The most common type of numerical change is *polyploidy*, where genome sets are multiplied. Polyploidy is regarded as a powerful force in plant evolution. It has been estimated that almost 70% angiosperms, including several cereals and vegetables, are polyploids. In pteridophytes, nearly 90% of plants are polyploid.

Polyploids, leading to diversity, may involve direct multiplication, viz. *autopolyploidy* or multiplication of different genomes in combination, viz. *allopolyploidy*. *Segmental polyploids* involve duplication associated with segmental changes.

Natural autopolyploids are rather rare. The species *Galax aphylla* and *Achlys triphylla* are considered to be natural autotetraploids, where not much structural changes have taken place in evolution. Similar autopolyploidy has also been recorded in *Allium tuberosum* where all the 32 chromosomes form quadrivalents (Sen, 1974a).

Of all higher plants that are polyploids, most belong to the very heterogenous group of allopolyploids, and the autopolyploid status has been evidenced so far only for a small number of species. *Segmental polyploids* occupy an intermediate position between auto- and allopolyploids. The proportion of polyploids is specially high in Gramineae whereas it is very low in gymnosperms.

Amphidiploids arising out of chromosome doubling in a sterile hybrid are very common in nature, classical examples being provided by *Nicotiana tabacum*, *Primula kewensis*, *Galeopsis tetrahit*, and *Biscutella laevigata*. Intergeneric crosses, followed by induced amphidiploidy at an intergeneric level, were made possible, leading to the formation of hybrid *Raphanobrassica*.

Müntzing raised a synthetic *Galeopsis tetrahit* originating from two/three species. The F_1, though sterile, had a triploid, which when crossed with *G. pubescens* led to the production of tetraploid *G. tetrahit* plant with $2n = 32$ chromosomes. Of the crop species, the best example of an autohexaploid is wheat, *Triticum vulgaris*, where three genomes AA BB DD are involved in its origin. AABB genome was obtained from *T. dicoccum* and DD genome from *Aegilops squarrosa*.

Two other classical examples of tetraploids are *Nicotiana tabacum* with $2n = 48$ chromosomes derived from *N. sylvestris* ($2n = 24$) and *N. tomentosiformis* ($2n = 24$). Of the genus *Brassica*, *B. napus* is an allopolyploid of *B. campestris* (mustard) with $2n = 20$ and *B. oleracea* (cabbage) with $2n = 18$ chromosomes. Besides these classical examples, amphidiploids are rather common in nature, including *Oryza sativa* (rice). Some other examples are *Primula kewensis* (containing genome of *P. floribunda* and *P. verticillata*) and *Dactylis glomerata*.

Synthetic *Triticale* involving two genera, *Triticum* and *Secale*, containing 42 chromosomes of wheat and 14 chromosomes of rye is well established.

The application of colchicine, introduced into mutation research in the 1930s, is still the most frequently used method for increasing the chromosome number and induction of polyploidy.

Fertility of experimentally produced autopolyploids

The transition from $2n$ to $4n$ is, in general, combined with a reduction of the fertility. This is the main reason for the reduced selection value of the autotetraploids in all those cases in which the species studied propagate sexually.

The degree of reduction of fertility varies widely in different species. With regard to seed fertility, the reaction of autotetraploids is rather negative. Pollen fertility is mainly governed by the course of microsporogenesis. Seed fertility, however, is a very complex phenomenon depending on several factors, including the degree of anomalies during micro- and megasporogenesis. Low seed production is often the result of a total effect of several negative factors. A diploid tomato plant, for instance, gives about 40 fruits, with 40–100 seeds per fruit. An autotetraploid plant, on the other hand, has only four fruits, with about 10 seeds each.

Meiosis in autotetraploids

In the pollen mother cells of an autotetraploid plant, each chromosome of the complement is present in the form of four homologues. They show a characteristic pairing behavior: either one quadrivalent or two bivalents are formed during zygotene and pachytene. It was already demonstrated by Darlington (1937) that close pairing of homologous chromosomes, as observed in the PMCs of diploid plants, occurs principally between two chromosomal units, irrespective of the number of homologues present in the nucleus. This process is called "primary pairing." In an autotetraploid nucleus, primary pairing can occur in two different modifications:

1. two normal bivalents arise, each showing close chromosome pairing at pachytene or
2. the four homologous chromosomes are united to a cross-shaped configuration, a quadrivalent, which shows primary pairing in its four arms.

As the four chromosomes of each group are fully homologous and structurally identical, they can replace each other during the course of pairing.

Spatial location of the four homologous chromosomes within the nucleus seems to be responsible for different configurations due to pairing. If they are distant from each other, the only probability is that they

are united in single pairing configurations and as such two bivalents arise. If, however, they are closer, a tetravalent is formed. It is obviously a matter of chance, whether two bivalents or one quadrivalent are formed, depending on the details of the course of pairing. These procedures can be clearly analyzed at the pachytene of autotetraploid tomatoes.

If chiasmata formation takes place in each of the four arms of the quadrivalent, a ring of four chromosomes is formed in diakinesis and metaphase I, which is easily detected under the microscope. Lack of chiasma formation on one arm of the pachytene configuration results in a chain of four chromosomes; on several arms, it leads either to the formation of chain of three homologous chromosomes where the fourth one is present as a univalent, or to the formation of two rod-like open bivalents.

As postulated by Darlington, there is another mode of pairing in the PMCs, called "secondary pairing." This involves a very loose pairing between two homologous bivalents. In these cases, the four homologues of the group are present in the form of two normal bivalents which, however, show clear mutual relations with regard to their spatial orientation. The corresponding regions of the two bivalents lie more or less parallel within a short distance from each other, demonstrating thereby that they belong together.

The number of quadrivalents per PMC varies considerably not only between different species but also between different varieties of the same species. Examples are autotetraploids of *Lycopersicon esculentum*, *Asparagus officinalis*, *Oryza sativa*, *Secale cereale*, and many others. Very high values were found in $4n$ *Allysum maritinum* and *Glycine max*. Unusually low quadrivalent frequencies are known in autotetraploids of *Cuminum cyminum*, *Raphanus sativus*, *Lotus corniculatus*, *Vaccinium corymbosum*, among others. In some species, there is no quadrivalent formation at all in autotetraploid plants, as in *Solanum rybinii*, *Agathaea coelestis*, and *Tagetes erecta*.

In the PMCs of several autotetraploid plants, there is a normal distribution of the chromosomes from the ring of anaphase I, followed by an undisturbed continuation of the course of meiosis. Genomatically balanced diploid microspores arise, giving rise to fully functional diploid gametes. Misdistribution from the rings to chains, especially the presence of tri- and univalents, however, leads to the formation of unbalanced gametes. They cause the occurrence of aneuploid plants in the progenies of the autotetraploid, which reduce seed production of the tetraploid strain as a whole. Their frequency can be very high, almost 50% in $4n$ *Sesamum annum* and *Lolium perenne*.

In a small number of tetraploid PMCs, the course of meiosis is altered in such a way that triploid and haploid microspores are formed instead of the diploid ones. In self-propagating species, diploid plants can arise in the offspring of tetraploids due to the union of double reduced male and female cells.

Meiosis in Allopolyploids

The other type of polyploidy, which is rather more prevalent, is the duplication of chromosomes of different sets before or after the attainment of hybridity. Such allopolyploids, having different sets of chromosomes duplicated have been involved in the evolution of different crop species including rice, wheat, and cotton. As each genome in an allopolyploid is present in duplicate, bivalent formation is ensured and the resultant plant is often termed *amphidiploid*. In the evolution of plant species in an allopolyploid, often the different genomes may have partial homology with each other, the term applied being "segmental polyploid", which often shows multivalents.

Meiosis in a hybrid, as expected, is bound to be irregular due to nonhomology of parents. The irregularity leads to formation of univalents, irregular separation of chromosomes, and formation of unbalanced gametes leading to sterility. To overcome the problem of sterility and ensure bivalent formation in meiosis, doubling of chromosomes either prior to the formation of gametes or after zygotic formation from gametes is resorted to. The most common chemical applied for the artificial doubling is colchicine, the alkaloid from the plant *Colchicum luteum*. Production of autotetraploid or chromosome doubling of hybrids is done to induce allopolyploids and amphidiploids in order to ensure smooth pairing of chromosomes and seed fertility. Classical cases of natural and induced amphidiploids include *Raphanobrassica*, an intergeneric hybrid raised by G.D. Karpachenko in 1927, resulting from crosses between *Raphanus sativus* ($2n = 18$) and *Brassica oleracea* ($2n = 18$) followed by chromosome doubling. The sterile hybrid showed 18 chromosomes where the fertile amphidiploid had 36 chromosomes.

Primula kewensis is also an amphidiploid resulting from a cross between *P. floribunda* and *P. vertcillata*, both containing $2n = 18$ chromosomes; the induced amphidiploid is characterised by $2n = 36$ chromosomes.

A classical case of interspecific amphidiploid is *Galeopsis tetrahit*, synthesized by A. Müntzing by crossing *G. pubescens* and *G. speciosus*, both containing $2n = 16$ chromosomes, the amphidiploid *G. tetrahit* having $2n = 32$ chromosomes.

The best example of an amphidiploid crop is wheat, where three different genomes are involved. The common wheat is hexaploid with $2n = 42$ chromosomes. The different genomes involved in the ancestry of wheat are those of *Triticum aegilopoides*, *Aegilops speltoides*, and *A. squarrosa*, all containing $2n = 14$ chromosomes. The three genomes are designated as A, B, and D, respectively, though the donor of the B genome is still disputed. As all three species belonging to *Triticum* complex do not differ widely from each other, hexaploid wheat often shows chromosomal homology.

Other important examples of amphidiploids or allotetraploids are *Gossypium hirsutum* and *Triticale*, the former being termed as upland cotton. Both the Old World and American cotton have $n = 13$ chromosomes, the former having large and the latter having small chromosomes. New World cotton has 26 pairs of chromosomes, of which 13 pairs are large and 13 pairs are small. Thus, New World cotton is the stable type, and the cultivated type is the product of both Old World and upland cotton. This ancestry was confirmed by J.O. Beasley, crossing two diploid species *G. herbaceum* with $2n = 26$ and *G. raimondi* with $2n = 26$ chromosomes. Tetraploid cultivated cotton is *G. hirsutum* with $2n = 52$ chromosomes.

The genus *Triticale* is a cross between wheat and rye, where the ancestors involved are *T. aestivum* ($2n = 42$) and *Secale cereale* ($2n = 14$) chromosomes. The hybrid is sterile with 28 chromosomes. Doubling of hybrid chromosomes led to the formation of octaploid *Triticale* with $2n = 56$ chromosomes (vide Gupta and Priyadarshan, 1980).

Haploidy

In flowering plants, diplophase dominates; the haplophase is normally limited to the pollen grains and embryo sacs. In exceptional cases, plants that are entirely haploid may arise. With regard to all their parts, haploids are smaller and often display poor vigor.

Poor vigor applies particularly to haploids of cross-fertilizers, such as rye and timothy, in which the general occurrence of recessive, deleterious genes in the populations often affects the haploids. In such a plant, there is only one set of chromosomes and a recessive defect, therefore, immediately is expressed.

Meiosis in a true haploid is, of course, very irregular. Chromosome pairs consisting of two homologous chromosomes united by chiasmata cannot be formed, and as such, chromosomes are distributed at random: some of them move to one pole, others to the opposite pole. Often, however, a low frequency of functional gametes is formed; these gametes are unreduced.

In *polyhaploids* (i.e. haploids obtained from polyploid species), for instance wheat group which is hexaploid with $2n = 42$, conditions will be more complicated because haploids will contain more than one genome. If the genomes are homologous or partially homologous, bivalent formation and good fertility in the polyploid may be the consequences.

Haploids may arise spontaneously, but as a rule, in only very low frequencies. In certain induced auto- and allopolyploids with high chromosome number, the tendency to halving the chromosome number may be more pronounced. This is probably because the chromosome halving in such cases implies a reversal of the chromosome number from an abnormally high level back to the normal condition.

Delayed pollination too may result in haploidy. In this method, anthers are first removed and pollination is delayed until the egg cell starts to divide without having been fertilized. At any rate, the haploid plant arises because a haploid cell nucleus starts to divide and gives rise to the embryo. The other method is to use X-rayed pollen. It has been possible, especially in einkorn wheat *Triticum monococcum*, to obtain haploids in fairly high frequencies, using these methods.

Instead of X-rayed pollen, it has been found that pollen of different species may also be used with advantage, representing a third method. If, for instance, *Solanum nigrum* is pollinated with spores from *S. luteum*, haploids of *S. nigrum* may arise; these haploids are derived from unfertilized egg cells of *S. nigrum*. A fourth method to obtain haploids is through *twin method* and *polyembryony*. From the same seed, two plants may grow up, which can be separated and survive as independent individuals. These plants are usually derived from two ovules from different embryo sacs. It has been found that one of the twins in such a pair sometimes has a deviating chromosome number and may give rise to a triploid. Moreover, haploids frequently arise as twin plants.

The most convenient method for induction of haploidy is anther and pollen culture. The anther with pollen grain having one set of chromosomes can germinate in vitro to give a haploid plant, as done in *Datura* and later in several crop species. The haploids thus produced can serve as good materials for production of homozygous diploids through colchicine treatment.

Aneuploidy

In addition to polyploidy, where multiplication of the entire genome is involved, biodiversity can also be generated through aneuploidy, where multiplication or reduction does not involve all chromosomes of the set. Aneuploids may be trisomic ($2n + 1$), tetrasomic ($2n + 2$), monosomic ($2n - 1$), or nullisomic.

Trisomy ($2n + 1$) has been known in the plant kingdom for a long time, but its modification could be worked out only in later years. In addition to *primary trisomy*, with three chromosomes of identical gene constitution, secondary and tertiary trisomics are also recorded. In *secondary trisomics*, the third chromosome is an isochromosome, whereas in *tertiary trisomics*, it is a translocation chromosome. All of them can be identified on the basis of breeding behavior, marker gene, and meiotic configuration. Primary trisomics form a trivalent of different configurations, secondary trisomics a ring of three or a chain, and tertiary trisomics often show a pentavalent configuration. Primary trisomics have been analyzed thoroughly in *Datura stramonium* by Avery et al.

(1959) and in *Lycopersicum esculentum* by Rick and Burton (1954). Trisomic embryos, as expected, do not mostly stand in competition with diploid ones. Primary trisomics, as originating from a triploid progeny or from asynaptic mutants, often serve as a tool in marking genes in specific chromosomes as well as in detecting gene expression. Trisomic mutants have been obtained in several crop species including jute (Iyer, 1968). Secondary trisomics have also been found in *Datura stramonium* and the tomato (Khush and Rick, 1964). Similarly, tertiary trisomics have been analyzed in tomato, where specific mutant genes in translocation chromosomes have been detected. Chromosomal analysis in this material is commonly carried out at the pachytene stage.

Monosomics

Origin and use of monosomics

The monosomics and nullisomics can be achieved through a variety of procedures.

From haploids. Monosomics can be produced by crossing a haploid with an amphidiploid. In wheat, Sears obtained an accidental production of two haploids in the progeny of crosses between rye (*Secale cereale, n = 7*) and a variety of hexaploid wheat (*Triticum aestivum*) known as Chinese spring. A haploid as female parent was crossed to normal hexaploid wheat as male, and monosomics and other types were produced. These monosomics when selfed gave nullisomics and trisomics, and the latter on selfing gave tetrasomics.

From interspecific hybrids. Monosomics have been produced in some crops by interspecific hybridization, followed by a back cross with the polyploid parent as in *Nicotiana*, from which monosomics are intended to be isolated.

From partial asynapsis. Plants that show partial asynapsis during meiotic metaphase I, do not show perfect bivalent formation, but produce a variable number of univalents. These univalents segregate to the two poles at random during anaphase I and deficient gametes with a missing chromosome are often produced. Such methods have been used in tobacco and wheat for producing monosomics. In wheat, the entire set of 21 monosomics has been obtained using haploids and partial asynapsis.

Irradiation treatment. It has also been used as an effective means for producing monosomics as in cotton (*Gossypium hirsutum 2n = 32*), where the inflorescence was irradiated. Monosomics are produced due to chromosomal aberration and deficiency in one pole, arising out of nondisjunction.

Alterations in Chromosome Structure

When both genomes of an individual are identical, without any structural changes inherited from either parent, the individual is a *homozygote*. If, however, a structural change is present in even one chromosome but not in its homolog, the complement is *heterozygous* for the change. Alternatively, both homologous chromosomes may carry the same alteration, making the complement *homozygous* for the alteration. Homozygotes form bivalents during meiosis while heterozygotes, depending on the nature of the

change, exhibit different types of chromosomal behavior. Heterozygotes, involving
minor alterations or alterations of heterochromatic segments, have a greater
chance of survival since there will, in
general, be less gross alteration in the
phenotype.

The alteration of a chromosome segment may, according to its size, affect
several genes, and thereby several sets of
phenotypic characters. Such changes in
chromosome structure are caused by two
events – breakage and reunion – according
to the classical theory. All breaks are
essentially similar, and fractures involve
the chemical bonds of nucleic acid and
protein. These regions are unstable or
labile sites where primary lesions are
formed. An interaction between two lesion
sites caused by the primary events leads to
exchange initiation.

Different types of alterations

Different types of structural aberrations
are (1) deletion and deficiency, (2) duplication, (3) inversion, and (4) translocation–reciprocal and nonreciprocal (Figs.
4.1, 4.2). The last two types are *qualitative*,
where only structural rearrangements
take place, the total amount of chromatin
matter or genes remaining the same. The
fate of the chromosome bearing the
alteration depends on the position of the
centromere and the amount of heterochromatin present.

Deficiency and deletion

When a chromosome loses a segment at
one end, the phenomenon is known as
deficiency and the chromosome is
regarded as a *deficient* one. Loss of an
intercalary segment is known as *deletion*.

Deletions are much more frequent than
deficiency. *Amphiplasty* or the loss of a
terminal satellite by breakage, across the
satellite stalk, and attenuation of the
chromosome is a form of deficiency
observed in certain plants, which does not
fit in with the telomere hypothesis.

The fate of a deficient chromosome
depends on the position of the centromere
and the size of the portion deleted. The
segment with the centromere, if it contains
most of the major genes, survives but the
fragment without the centromere (*acentric
fragment*) is lost. In plants with a diffuse
centromere (holocentric chromosome),
such as *Luzula*, fragments appear to
survive in succeeding generations.

Survival depends on the size and nature
of the segment lost. Individuals heterozygous for a small deletion or deficiency
may survive since the other homolog
carries the full complement of genes. The
loss of heterochromatic segments may be
compensated. However, homozygotes for
deficiency and deletion have very little
chance of survival.

Duplication

The presence of an extra segment of a
chromosome of a normal complement is
known as duplication. The segment may
form part of a member of a normal
genome, being present in *tandem* (i.e. next
to an identical segment) or in *reverse
tandem* (i.e. next to an identical segment in
a reversed position or in some other part of
the same chromosome). It may, in rare
cases, also exist as an extra chromosome.

Duplications may arise through (1)
primary structural changes of chromosomes; (2) unequal crossing-over of
chromatids or two homologous chromo-

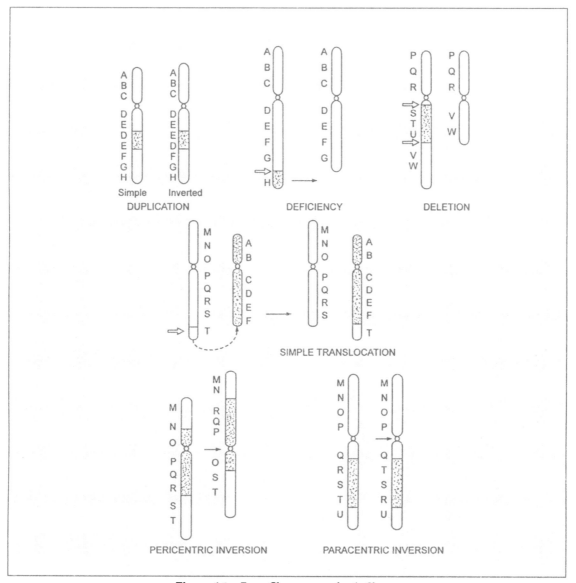

Figure 4.1 *From Chromosomes by A. Sharma*

somes. In the latter case, at the point of contact, two breaks occur, followed by possible recombinations of the four chromosome ends. The result is the formation of two new chromosomes by crossing over in inversion or translocation heterozygotes.

Duplications are observed more frequently in nature in the heterozygous condition, and are less likely to be lethal to an individual than deficiencies are, due to possible dosage compensation through repression of the duplicated genome. During meiosis, while pairing with a

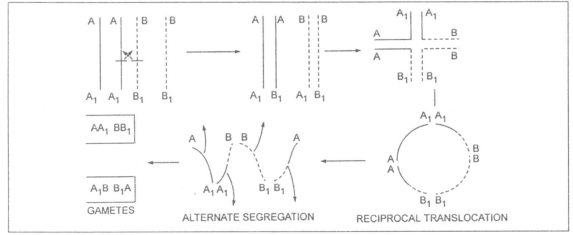

Figure 4.2   *From Chromosomes by A. Sharma*

normal homologous chromosome, the duplicated segment may bulge out as a loop. Homozygotes, however, have little chance for survival unless the segment involved is very small or totally repressed.

Inversion

The structural alteration occurring most frequently in higher organisms in the wild state and used widely in genetical experiments is *inversion*. It indicates the reversal in position of a chromosome segment. It is almost always intercalary, rarely terminal. The formation of inversion involves two breaks in a single chromosome at the point of contact between chromosome strands within a loop.

The primary effect of inversion is *position effect* or a change in the relative positions of genes. It results in phenotypic changes, sometimes involving variegation, as in leaf.

Meiotic behavior of a chromosome carrying an inversion depends on the position of the centromere in relation to the inverted portion.

In a *paracentric* inversion, only one chromosome arm is involved. There is no change in the shape of the chromosome concerned. This type is more frequent than the other.

A *pericentric* inversion includes the segment bearing the centromere. It may be *symmetrical*, in which the centromere is located close to the middle of the inverted segment, or *asymmetrical*, in which the centromere is not in the middle. The latter type results in a change in the chromosome morphology and can be easily detected.

The possible configurations during meiosis include the following:

1. Both the chromosomes with the inversion and its normal homolog do not pair, but remain as univalents giving rise to fertile gametes.
2. The two homologs pair along their homologous segments, the inverted part and the region homologous to it remaining as two opposite loops. Disjunction will result in fertile gametes, with 50% bearing the chromosome with the inversion and 50% the normal one.
3. The inverted segment may loop over itself to pair with its homolo-

gous counterpart in the normal chromosome. In a paracentric inversion, a crossing-over may take place between the two chromosomes within the loop. When the chromosomes separate, one chromosome will have two centromeres (dicentric), while the other will be acentric. During anaphase, the two centromeres of the dicentric chromosome tend to move to opposite poles, giving rise to a typical *dicentric inversion bridge*. The acentric chromosome remains at the equator or may divide and two acentric fragments may be observed at the two poles. The bridge may break in the middle, the two broken pieces moving to different poles or the entire bridge may be lost during cell division. In rare cases, as for example in species of *Triticum* reported by Sears, the entire bridge may move to one pole and be maintained as such through successive cell divisions. However, in all cases, after the formation of the inversion bridge, both the bridge and the fragment are ultimately eliminated. If included within a gamete, it is sterile. Double fragments and double bridges may arise if all the four chromatids are involved in the crossing-over. If the bridge remains at the equator, 50% of the gametes are nonviable, and if included within one daughter cell, that gamete becomes nonviable.

When the inversion is *pericentric* and the centromere is included within the loop, and crossing-over occurs, each resultant chromosome has a centromere. However,

the compositions of both the daughter chromosomes become altered and therefore the inversion bridge is not formed. In the changed segment, some genes are duplicated, and others are deleted. The phenotypic effects are accordingly altered. Occurrence of more than a single inversion in a chromosome is known as *complex* or *multiple* inversion.

Shift denotes the transfer of an intercalary segment to another position within the same chromosome, either to the same or to another arm. It involves three breaks and rejoinings. If the segment is a small one, the chromosome, during meiosis, will pair with its normal homolog, the nonhomologous portions bulging out as loops. When the segment is placed in an inverted position, it will behave as an inversion.

Structural heterozygosity involving inversion and heteromorphicity in chromosome pairs is also not uncommon in plants. Cryptic structural hybridity, involving only minute segments, has been worked out in several plant genera such as *Oryza*, *Corchorus*, *Trillium*, *Plantago*, *Fritillaria*, *Alocasia*, *Allium*, and *Sida* (Sharma and Sharma, 1959; Stebbins, 1971). Intrachromosomal structural changes detected through genetic studies were originally interpreted in various ways. Inversion effects were mostly explained as due to crossover suppressor genes, and deletion effects as due to gene mutation. The visible effects of deletion in heteromorphicity or loops and inversions resulting in dicentric bridges made cytological interpretation more precise. Inversions often result in nonviability of the gamete due to meiotic irregularity. McClintock (1941) worked out the *breakage–fusion–bridge cycle* in maize, suggesting the perpetuation of such aberrant chromosomes through different cell cycles. The breakage of the dicentric bridge at anaphase results in deficient chromosomes with broken ends in daughter nuclei. The pollen grain nucleus may develop, provided the acentric compensating fragment is included. The two deficient chromosomes in two sperm nuclei may even enter into the formation of zygote and endosperm, the viability of the latter depending on the nature of the deficiency gene. McClintock (1950) worked on special mutable loci (Ac–Ds) in maize controlling breakage. Its details have been dealt with under mobile sequences (Chapter 1 of this book). Evidence of inversion heterozygosity has been obtained through refinements in methodology particularly in pachytene chromosomes. The occurrence of such inversion in nature shows the extent to which it plays specific roles in natural selection and evolution of species. The restriction of crossing-over in certain segments thus helps to maintain certain desirable combinations and shows the adaptive value of inversions. Heteromorphicity in chromosome pairs arising out of fragmentation and translocation of chromosome segments has thus played comparatively little role in evolution. Such heteromorphicity in chromosomes arising out of deletion and translocation has been recorded in *Cipura paludosa*, *Zebrina pendula*, and *Taraxacum officinale* (Sharma, 1974). However, in species reproducing through vegetative means, such structural changes play a very effective role in evolution through their survival in somatic tissue.

Translocations

Variability of chromosome structure arising out of chromosomal aberrations may be interchromosomal or intrachromosomal. Of the interchromosomal aberrations, *translocation*, which mostly involves normal interchange of segments as in crossing-over, is quite common in plants.

Several types of translocations have been recorded, depending on the number of breaks and rejoining: (1) simple translocation; (2) reciprocal translocation; (3) shift within the same chromosome; (4) insertion in to a different chromosome, between nonhomologous or homologous chromosomes; and (5) centric fusion (Fig. 4.3).

Simple translocation. In a simple translocation, a terminal segment from one chromosome may get broken off and attached to one end of a nonhomologous chromosome. It has been reported in some monocotyledonous plants and is possibly facilitated by presence of telomeric heterochromatin.

Reciprocal translocation. It involves the exchange of segments between two nonhomologous chromosomes, without any loss or gain of the total chromatin matter. This exchange takes place through two breaks on two nonhomologous chromosomes (AA_1 and BB_1) and subsequent rejoining of the broken ends of one to the corresponding broken ends of the other to give two translocated chromosomes AB_1 and BA_1 (Fig. 4.2).

Two types of such mutual translocations are possible. In one, two new monocentric chromosomes are formed (*symmetrical interchange*) while in the other, a dicentric and an acentric chromosomes arise (*asymmetrical interchange*), which are not transmissible. In cases of *symmetrical interchange*, when the two translocated chromosomes (AB_1 and BA_1)

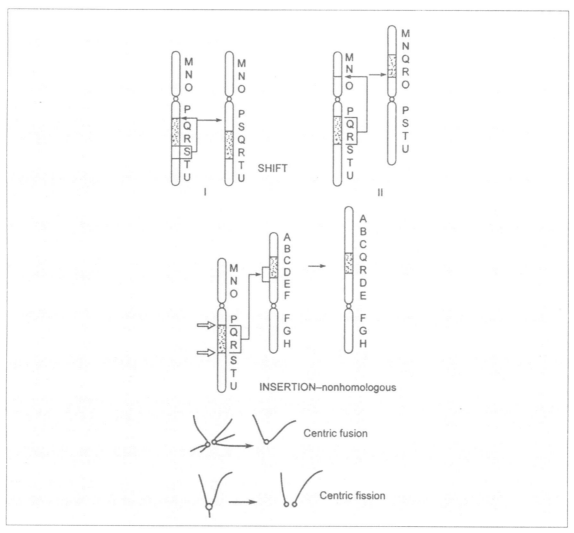

Figure 4.3 Complex translocation may involve three or more breaks.
From Chromosomes by A. Sharma

undergo pairing with their normal homologues (AA_1 and BB_1) during meiosis, the four chromosomes pair along their homologous segments. A cross is formed at first, which is converted on the terminalization of the chiasma to give a transloca-tion ring. Later, the ring opens out into a chain.

Three types of combinations are possible:

$AA_1BB_1 \times A_1BB_1A$: a normal and a translocated chromosomes produced by alternate chromosomes of the chain going to the same pole.

$AA_1A_1B \times BB_1B_1A$ and $A_1B\, BB_1 \times B_1A\, AA_1$ both produced by adjacent chromosomes going to the same pole.

In the two latter cases, duplications and deletions are present and the combinations are lethal.

Therefore, the only possible viable configuration is the first one, AA_1BB_1 and A_1BB_1A in which alternate chromosomes have moved to the same pole. Thus, of the two sets of viable gametes formed, one bears the original chromosome AA_1BB_1 and the other the translocated one AB_1BA_1.

From these two gametes, with the original configuration (O) and the changed configuration (C), three types of combinations are expected in the offspring, namely,

O × O: original, homozygous for the original complement, bearing both unchanged chromosomes. During meiosis, such individuals form bivalents.

C × C: homozygous for the changed complement, both chromosomes being of the changed type. During meiosis, only bivalents are formed.

O × C: heterozygous, in which one chromosome is unchanged and the other changed. These individuals show behavior similar to their parents; the chromosomes form translocation rings, undergo alternate segregation and give rise to three types of progeny as described before. A translocation involving three chromosomes will give a ring of 6, or 4 chromosomes a ring of 8 and so on.

In *Datura stramonium*, all the three types of progeny, O × O, O × C, and C × C have been observed.

In certain genera such as *Oenothera* ($n = 7$), one species *O. hookeri* has 7 bivalents and the chromosomes are unchanged (O × O). Other species such as *O. biennis* show the progressive formation of rings of 4, 6,

8, and 10 chromosomes, giving a series of C × O configurations where 2, 3, 4 and 5 chromosomes have undergone reciprocal translocations. Finally, in *O. lamarckiana*, a ring of 14 chromosomes is observed, suggesting that all 7 chromosomes have entered into reciprocal translocations with each other. The ring of 14, or of 12 and one bivalent, later opens out in a chain. Adjacent chromosomes go to opposite poles during anaphase, forming two sets of 7 chromosomes, named *gaudens* and *velans* which remain constant through successive generations. The entire complex is named the *Renner complex* after its author. This type of translocation heterozygote in which only the combination C × O is possible is known as a *balanced heterozygote*. Any other combination homozygous for either *gaudens* (G × G) or *velans* (V × V), is not viable either due to *heterogamy* – the relatively better sponsoring of the velans complement in the ovule and the gaudens in the pollen mother cell – or due to a gene causing *balanced lethality*, which is lethal in deficient or duplicate doses.

Directed alternate segregation also depends on the length of the chromosome, position of breakpoints, number and position of chiasma, degree of terminalization, and location of centromere.

The monotypic genus *Rhoeo discolor* of the family Commelinaceae is an ultimate type of balanced heterozygote in which the intermediate stages have been eliminated. All its 12 chromosomes form a ring and undergo alternate segregation during anaphase.

Burnham (1962) prepared an extensive review on translocation interchanges in plants. This behavior has not been as effectively utilized as possible in utilitarian research, and a thorough analysis may ultimately lead to its effective exploitation. In addition to the classical case of *Oenothera lamarckiana* with Renner Complex (gaudens and velans), several other translocation heterozygotes have been reported from time to time in *Datura, Rhoeo, Hordeum, Collinsia, Zea*, and others (Bloom and Lewis, 1972; Garber, 1972). Of the three types of progeny expected out of translocation heterozygotes, species such as *Rhoeo discolor* and *Oenothera lamarckiana* represent cases of balanced heterozygosity, where the heterozygotes survive whereas the other two homozygotes originating out of alternate disjunction are lethal. In maize, even supernumerary B chromosomes have been found to be involved in reciprocal translocation. Several linkage groups have also been assigned to respective chromosomes. Translocation heterozygotes, through their meiotic behavior, often serve in identifying chromosome ends and their genetic constitution.

Regarding translocation, the following facts may also be taken into account.

Following polyploidy too, considerable amount of structural rearrangements takes place in evolution. These structural changes, occurring at an intergenomic level, are often translocations as in *Nicotiana tabacum* (2*n* = 48), *Avena maroccana* (2*n* = 28), and others.

Translocations at an intergenomic level in polyploids, have been classified under two categories:

1. *random* translocations,
2. *species-specific* translocation (Jiang et al., 1994).

The former occurs in different chromosomes, in different populations, or in some polyploid species. On the other hand,

species-specific translocations involve only specific chromosomes and are found in every population. Species-specific translocations have been accounted for through a *"nucleo-cytoplasmic interaction hypothesis"* of genome evolution (Gill et al., 1991). According to this theory, a new amphidiploid must pass through a limitation of sterility, arising out of adverse interaction between the male nuclear genome, and nucleus and cytoplasm of the female parent. Therefore, to restore fertility and nucleocytoplasmic compatibility, certain specific chromosome changes must occur. Through the translocations, the balance of fertility is restored. Intergenomic translocations have been observed in the tetraploid wheat, both in *Triticum timopheevii* and *T. turgidum*. In the former, it is random, whereas in the latter, species-specific intergenomic translocations are involved.

In case of *Nicotiana tabacum*, where the different genotypes are involved, several homozygous translocation lines have been studied. It was recorded that one recombinant is common to all genotypes, which represents species-specific chromosome translocations. Its presence, following polyploidy, is necessary for restoring nucleocytoplasmic compatibility.

Insertion involves the transfer of an intercalary segment to an intercalary position in another chromosome, which may be homologous or nonhomologous with the original chromosome. If inserted within a homologous chromosome, the original one will be deficient and the one with the insertion will bear a duplication for the segment concerned. Meiotic behavior of the two chromosomes will be modified accordingly. When the insertion is in a nonhomologous chromosome, the behavior will depend on the size of the segment, which may be identified by banding.

Centric fusion or Robertsonian translocation. This involves reciprocal translocation across the centromeric regions of acrocentric chromosomes or the fusion of telocentric chromosomes to form a metacentric chromosome. It is a special type, not strictly involving breakage, and has been observed principally in mammalian chromosomes.

The other possibility with regard to centric fusion includes both breaks very close to the centromere, one on the long and the other on the short arm so that rejoining gives rise to a large metacentric and a very small chromosome. The latter is lost in the course of a few cell generations. Alternatively, both breaks may be on the short arms, leading to a metacentric with two centromeres, so close together that they may function as one, and an acentric chromosome. Other tandem type fusions are (1) fusion of two acrocentrics to give a double length acrocentric with loss of a centromere or (2) fusion of an acrocentric, after loss of its centromere, with a metacentric.

Centric fusion gives some clearly observable results, such as (1) loss of short arms, (2) formation of a new linkage group: the large metacentric, (3) reduction in the total number of linkage groups, and (4) in some cases a modification of chiasma frequency and location in the derived chromosome.

The reverse process of centric fusion, that is *centric fission* or *dissociation*, assumes the transformation of a metacentric chromosome into two acrocentrics in the course of evolution (Fig. 4.3).

According to the telomere concept, both centric fusion and dissociation are special types of mutual translocations involving the loss and the gain respectively of one centromere and two telomeres. These two processes can increase or reduce the chromosome number without a change in the number of long arms, according to Robertson's law that the number of chromosomes may vary but the number of arms remains constant. They have been found to be responsible for polymorphism in different species.

Position effect

Position effect is observed when genes or chromosome segments are transferred to a new position and this phenomenon alters the phenotype. Of the two types of position effect commonly known, stable (S) and variable (V), the former is caused by duplication in the euchromatic segment. A well-known example is maize observed by McClintock in 1952. If a wild type gene is transferred to a heterochromatic segment, the activity of the gene is repressed. The degree of repression is variable and affected by environmental factors, leading to somatic mosaicism and variegation. A classical example is the *activator-dissociation system (Ac–Ds)* in maize, which depends on the action of two separate loci, the *Ac* and the *Ds*. The latter only functions in the presence of the former. Simultaneous occurrence of both loci increases the frequency of chromosome breakage and subsequent formation of different types of chromosomal aberration. *Ac* and *Ds* are visualized as blocks of heterochromatin, which can move to different sites of the chromosome complement by *transposition*. The amount of heterochromatin involved may also change. Increasing dosage of *Ac* delays the expression of *Ds* proportionately.

Ring chromosomes

The term is usually applied to a deletion ring in which the centromere is in the middle of a coil, the terminal segments break off, and the broken ends of the coil rejoin. Other rings observed during meiosis are rings formed by large bivalents opening out at diakinesis; by multivalents, when the chiasma are terminalized; by translocation heterozygotes, as described earlier and by isochromosomes. All these rings can be distinguished on observing the subsequent stages of division, particularly anaphasic segregation.

Some centric rings formed by deletion are stable. More commonly, they tend to change in size or even be lost. During mitosis, the duplicated ring chromatids may move to opposite poles to be included in daughter cells; they may form a double sized ring or an interlocking ring.

Further division results in the following: (a) The interlocking ring may break, segregate regularly and fuse; (b) A dicentric ring may break asymmetrically and the resultant chromosomes will carry the duplication–deficiency complements; (c) The ring may be lost by anaphase lagging; or (d) The ring may be included in either pole through nondisjunction.

The subsequent cell generations may carry complements varying between monosomy and tripolysomy, with variable duplication–deficiency phenotypes.

In higher organisms, the ring chromosomes are ultimately lost in subsequent cell divisions, partly due to the difficulties in their transmission. Small rings, as in maize, can separate easily after replication while large rings are lost earlier.

Fragmentation

The importance of fragmentation in the evolution of species needs elucidation. Fragments may be spontaneous or induced (Sharma and Sharma, 1960) through different physical and chemical agents, including organic compounds such as plant pigments (Sharma and Gupta, 1959) and vegetable oils (Swaminathan and Natarajan, 1957). Under normal conditions, the fragments, if not translocated, undergo deletion and, as such, their evolutionary possibilities become limited. However, centric fragments are liable to be maintained, in which case their scope in evolution is indicated. Such examples are common in species of *Scilla* and several other genera. In Musaceae and Cucurbitaceae, fragmentation of chromosomes across secondary constriction regions has been suggested to be a physical basis of speciation (Bhaduri and Bose, 1947; Chakravarty, 1951). The theory involves the origin of centromere *de novo*, for which evidence has yet to be obtained.

Fragmentation of chromosomes may result in an increase in chromosome number such as in species of *Luzula* where the chromosomes are either polycentric or possess diffuse centromeres (Camara, 1957). The change in phenotypic character associated with fragmentation in such chromosomes is likely due to position effect since fragmentation merely involves grouping of genes in smaller units.

PART II

Manipulation and Transfer of Chromosomes and Genomes through Hybridization

Chromosome engineering had its inception in the pioneering contribution of Sears (1956), who utilized irradiation as a means to translocate a gene for leaf rust resistance from *Aegilops squarrosus* to wheat. The establishment of aneuploid lines permitted manipulation of individual chromosomes, and genetic analysis of wheat and its relatives. The production of a large number of interspecific hybrids and amphidiploids aided the generation of several addition and substitution lines in this group. Such lines proved to be of much help in transfer of foreign genes in this group of plants (Feldman, 1988; Sybenga, 1995).

In the primary pool of wheat, donors of all the three genomes, AA, BB and DD, are included. Transfer of genes from the primary pool could be achieved through hybridization, which permitted crossing over of homologous genes, and through back-crossing.

Closely related species of amphidiploid *Triticum* and *Aegilops*, having genome affinity with wheat, are included in the secondary pool. The latter also covers the Sitopsis section of *Aegilops* contributing possibly to B genome. The transfer from secondary pool is possible by hybridization, back-crossing, and selection of homologous genome. In wheat and its allies, homologous pairing and crossing-over can be induced genetically in interspecific and intergeneric hybrids. Transfer of alien segment through induced

homologous pairing can be achieved by hybridizing alien species, synthetic amphidiploids, or addition and substitution lines of certain genotypes, which allow homologous pairing. Several genes of common wheat, such as Phl, the pairing gene located in 5B which restricts pairing between homologs, can be deleted to achieve homologous pairing. The latter objective can also be fulfilled by counteracting the effect of Phl by high pairing gene of *Aegilops*, which allows several Triticineae chromosomes to pair with their wheat homolog.

Finally, both diploid and polyploid species, containing genomes nonhomologous to wheat, but related through possible homologous (partial homology between chromosomes of different genomes) segments, are included in the tertiary pool.

Of all special methods applied, *wide hybridization* coupled with *embryo rescue*, has been very successful. The technique involves hybridization between recipient and donor species, which are genomically widely different from one another, such as *Triticum, Zea, Elymus, Hordeum, Secale,* and others, in addition to the members of the secondary pool, with homologous and nonhomologous segments. In such cases, limited number of embryos can be formed. In wheat and maize hybrids, the embryo initially showed clear haploid complements of wheat and maize. But in successive cell cycles, all maize chromosomes were eliminated, leaving only one haploid set of wheat. The haploid embryo on culturing resulted in the production of a haploid plant. In crosses involving *Hordeum bulbosum* and wheat, partial elimination of barley chromosomes was noted, resulting in wheat–bulbosum addition lines. Similarly, wheat–maize lines, as well as oat and maize crosses were also obtained.

Monosomics have been extensively utilized by Sears in the chromosome substitution technique in *Triticum aestivum* and others, in which one or two chromosomes from any desirable species or variety can be introduced in another, replacing some of the chromosomes of the latter to achieve desirable combinations. In this technique, the normal donor genome homozygous for the desirable recessive is crossed with all the 21 monosomes of the recipient. The F_1 monosome hybrid is then back-crossed with the recurrent recipient and the respective monosome hybrid is tested and selected. The same process is resorted to several times, until a monosome can be obtained with only one desirable chromosome of the donor, the rest being bivalents of the recipient. By self-pollination, a fertile individual can be synthesized with 21 bivalents, of which one bivalent is of the donor. Extensive work has been done on wheat and its relatives (Morris and Sears, 1967), in which disease resistance and the other desirable characters have been combined with each other and all the 21 substitution lines have been obtained. The method also helps in mapping of genes in specific chromosomes.

A similar method has been employed to achieve the complete *substitution of the genome* in the cytoplasm of another. In this procedure, the species whose cytoplasm needs to be utilized, is kept in the female parent, and continued back-cross is performed with pollen parent of another species. It ultimately results in the complete replacement of the nucleus of one variety with that of another, often associated with male sterility. Genome substitution has been successfully secured in several genera such as *Triticum* and *Aegilops* (Sears, 1969).

Addition Lines

Amphidiploids permit the expression of foreign genes, avoid loss of chromosome integrity in addition lines, and serve as a source material for gene transfer. However, because of difficulty in securing addition lines from a basic alloplasmic source and to have it in a euplasmic background, the first back-cross involved an amphidiploid as the male parent.

There are certain limitations in securing full set of addition lines of one alien genome. The principal problems are genetic sterility associated with certain chromosomes, genocidal effects under certain combinations, and even poor pollen transmission in certain genotypes. It is possible to add alien chromosomes to alien addition lines, but because of the loss of genotypic balance, the genotype may ultimately lose the alien chromosome and return to euploid state. To avoid this difficulty, success was achieved in wheat by crossing the alien with a proper wheat monosomic, and recovering F_2 plants deficient for wheat pair and disomic for alien pair.

Substitution Lines

Substitution lines are those in which a chromosome or a complete genome of the alien species can be substituted in the target recipient.

Gene and Chromosome Translocations

To secure translocations of the alien gene, substitution or addition lines, as used in wheat, are intermediate steps. Several strategies have so far been adopted for securing translocations.

The first strategy is to use homologous pairing in substitution or addition lines. In wheat, this approach requires the elimination of 5B chromosomes, controlling pairing using pH1B mutant (Sears, 1981), or inhibiting the expression of the Phl gene. The Phl gene is obtained from *Aegilops speltoides*. The high-pairing gene (Chen et al., 1994) suppresses the effect of Phl and induces homologous pairing. Being dominant in single dosage, it can promote alien translocation without removing Phl. But it is not as effective as the removal of 5B in inducing homologous pairing (Jiang et al., 1994).

An indigenous method was worked out by Sears (1981), substituting an arm of one alien chromosome for the corresponding arm of wheat. The method takes advantage of the fact that univalents in monosomic wheat often misdivide at meiosis, leading to telocentric chromosomes. Similarly, univalent alien chromosome too may become telocentric by misdivision. The fusion of two telocentrics may result in a heterobrachial chromosome, one arm being of wheat and other of the alien chromosome. In view of the low probability and complicated procedure, this method is less used as compared to other methods of substitution. Such plants have stable constitution, as observed in Eastern European cultivar of wheat.

Chromosome substitution with foreign gene can also be obtained as a special strategy. The method involves a cross between a monotelosomic and a different diploid species to secure a nullisomic amphidiploid, followed by back-cross to a monotelosomic with a male nullisomic amphidiploid. The homologous partner of the diploid species will act as a substitute for the missing chromosome in the nullisomic amphidiploid.

The monotelosomic amphidiploid should be checked for the fertility or sterility of the male. Several substitution lines can be prepared using an amphidiploid for each line. The availability of amphidiploid and fertile nullisomic lines is obligatory for this approach.

Substitutions, including intervarietal transfer, require the availability of monosomics and nullisomics for all chromosomes, which are available in wheat, maize, *Nicotiana*, as well as cotton.

DNA Fingerprinting and Biodiversity

Each individual has a unique DNA pattern and is a component of the broad spectrum of biodiversity. The demonstration of this uniqueness on the basis of DNA sequence difference forms the basis of DNA fingerprinting. In other words, molecular evidence of biodiversity is DNA fingerprinting. Fingerprinting of DNA sequences in an individual is a molecular manifestation differentiating one individual from other. The term was originally coined by Jeffrey and his colleagues (1988), in relation to animal systems for simultaneous detection of variable DNA loci by hybridization of specific multilocus probes with restriction fragments. Digestion of DNA with specific enzymes, electrophoresis, blotting and detection, following hybridization with specific oligonucleotide probes, have made DNA fingerprinting a convenient technique. It is now widely applied in plant systems for detection of genetic diversity, genotype identification, in evolutionary studies, phenotype demonstration, and in molecular approach to biodiversity in its entirety.

There are two methods of fingerprinting, often termed as "DNA typing" or "DNA profiling" (Weising et al., 1995). In the first method, genomic DNA is cut with restriction enzymes, followed by electrophoresis and hybridization of the electrophoresed fragments with labeled multilocus probes, yielding a specific band pattern in the gel. The other method is to utilize the polymerase chain reaction, amplifying specific DNA sequences with oligonucleotide primers, electrophoretic separation of amplified sequences, and detection of polymorphism through staining of bands.

DNA fingerprinting based on hybridization is similar to the RFLP (restriction fragment length polymorphism) technique. But the RFLP technique is diallelic, involving two loci, whereas in fingerprinting, multiallelic or multilocus probes are used. The RFLPs are based on fragment length differences, obtained by digesting DNA samples with restriction endonucleases. Polymorphisms are due to the presence or absence of restriction sites in the genomes that are compared. The RFLPs have been used in gene mapping, preparation of phylogenetic tree, gene tagging, and establishing linked traits.

In fingerprinting, two types of multilocus probes are used: (1) *cloned DNA fragments* complementary to a minisatellite and (2) *microsatellites*. Both are tandem

repeats of oligonucleotides complementary to target DNA. The former is a basic motif of 10–60 base pairs; the latter is a simple sequence of 4–5 base pairs with low degree of repeats. A family of tandem repeated DNA sequences called minisatellites (Jeffrey et al., 1988; Weising et al., 1995) was noted to have a common GC-rich core sequence of 10–15 base pairs, in a tandem repeat of a 4 or 33 bp motif, in the intron of myoglobin gene. Hybridization of restriction fragments of human DNA with probes derived from the core sequences showed variable minisatellites in forming individual-specific DNA fingerprinting. In fact, the existence of variable number of tandem repeats formed the basis of polymorphism.

Both *mini-* and *microsatellites* occur at several genomic loci, often interspersed between unique sequences. Because copy numbers of both micro- and minisatellites with related motif are highly variable, these are grouped under the category of *"variable number of tandem repeats,"* otherwise termed as VNTRs, recorded in different groups of organisms including plants.

With the gradual refinements in methods, using synthetic oligomers, synthesized repeated sequences complementary to microsatellites have been successfully applied as multilocus probes in DNA fingerprinting. But as synthetic oligonucleotide probes are short and single stranded and can have very high polymorphic information, their use has greatly facilitated research on DNA fingerprinting. Information from such probes, complementary to short tandem repeats, is available on Data Bank. In addition to VNTR, random amplification of polymorphic DNA, termed RAPD, which may have a 10-base long chain, is also proving to be of much use in the search for variations.

Random amplified polymorphic DNA (RAPD) markers are obtained by PCR amplification of random DNA segments from single arbitrary primers. Williams et al. (1990) were the first to use the procedure on plant samples. The arbitrary primers used for the procedure are usually 9–10 bp in size; they have a GC content of 50–80%, and do not contain palindromic sequences. The number of DNA fragments that are amplified is dependent on the primer and genomic DNA used. A survey using DNA from 66 F_2 soybean plants indicated that RAPD markers are mostly dominant and heritable. Polymorphisms for RAPD may be due to single base pair changes, deletions of primer sites, insertions that increase the separation of primer sites over 3000-bp limit, and small insertions/deletions that result in changes in the size of the PCR product. When the PCR products were used as RFLP probes, they mapped at the same location as their PCR-amplified markers. Six out of eleven amplified probes tested with soybean DNA identified repetitive DNA sequences.

Like other molecular markers, RAPD can be used to tag chromosomes and genes, to fingerprint genomes, and to produce genomic maps. The advantages of using these markers are (1) a universal set of primers can be used for all species, (2) no probe libraries, radioactivity, southern transfers, or information on primer sequence are required, (3) the primer sequence is needed for information transfer, and (4) the process can be automated. The limitation of the use of RAPD markers is that they are dominant, which can be overcome by using more than one closely linked markers.

RAPD markers have been used to screen different strains of *Leptosphaeria maculans*, which is the causal agent of blackleg disease in crucifers. Differences between avirulent and virulent phenotypes, as well as among isolates within a pathotype, could be noted on the basis of specific RAPD markers. The results of RAPD analysis agree with previous classifications based on RFLPs and cultural assays and the question of ancestry of the fungus (*Leptosphaeria*) was also resolved. The main advantage of the RAPD analysis is its sensitivity and speed. A quick RAPD assay may be used to screen rapeseed seedlots for the presence of the virulent pathotype of fungus. The sensitivity of the method has been illustrated by the detection of polymorphisms among soybean (*Glycine max*) cultivars, with a single primer. This is a significant finding, as RFLPs indicated that soybean has a narrow genetic base and is highly homozygous.

Several short tandem repeats (STRs) have been recorded in a few plant species, also at mono-, tri- and tetranucleotide levels. In general, it has been recorded that in plant species, the most frequently observed class of STRs with AT sequences are ATA sequences. In general, in plants, short tandem repeats of 6 bp or more are quite common. Of the different types of STRs, tri- and tetranucleotides are abundant. The STRs of mono-, di-, tetranucleotide repeats occur in noncoding regions. Of the trinucleotide STRs, those containing GC base pairs are found in coding regions, wheres AT-rich ones are in noncoding regions. Interspersed STRs are quite common.

More than 50% of the trinucleotide repeats occur in coding regions of species of *Hordeum, Oryza, Triticum, Zea, Brassica, Glycine, Lycopersicum, Petunia, Pisum, Solanum,* and *Nicotiana.*

The number of STRs is also species-specific. In *Petunia*, it is one in every 11 kb; in *Brassica*, one in every 25.4 kb; and in *Arabidopsis, Zea,* and *Hordeum* it is one in every 42, 58, and 156 kb respectively. Rice has a variety of STRs, but only one type of STR is found in *Hordeum.* STRs occur more than two times frequently in dicotyledons (1 in every 31.2 kb) than in monocotyledons (1 in 64.5 kb). In dicots, 15% contain GC whereas in monocots 50% are with GC base pairs. Because of their high informativeness and ease of analysis, STRs in plants can be of much use in linkage mapping. STRs can be significantly useful in genetics, breeding, systematics, strain identification, and documentation of germplasm.

Both mini and simple sequence microsatellites are located in different chromosome loci, including telomeres and centromeres. Telomeric repeats located at chromosome ends are of a rather unusual nature. They are simple sequences with a clear function. There is marked base symmetry, one strand being GC-rich and other AT-rich, repeated several times, with a single strand 5' overhang in each chromosome. The end segment is synthesized by a special type of polymerase, *telomerase*, which utilizes RNA as the template.

PCR-based DNA fingerprinting takes advantage of in vitro amplification of DNA through thermostable enzymes and primers, both tandem repeat and arbitrary. The microsatellite core sequence as well as simple sequence oligonucleotide repeats, which were earlier used as hybridization probes, are now effectively employed as PCR primers for DNA fingerprinting (Lieckfeldt et al., 1993).

Moreover, for monitoring fidelity and molecular diversity, AFLP-amplified molecular fragment length polymorphism is also turning out to be a powerful tool in the study of biodiversity.

CHAPTER 5

Characteristics of Chromosome in Different Plant Groups

The plant kingdom represents a wide diversity of forms, ranging from unicellular alga to highly complex multicellular flowering plants. The habitats too differ widely, ranging from warm seabed to the desert and arctic zones of the mountains. This diverse milieu of growth and development, the morphological complexity of the widely different plant groups with their diversity of life cycle and physiology, are also reflected to a great extent in the genomic characteristics of distinct plant groups. It is better to discuss the characteristics chronologically of the groups according to their taxonomic hierarchy.

Algae

The algal system represents a very good example of evolution of complexity from a primitive structure. Leaving aside Cyanophyceae, with a primitive structure without any well-organized chromosome, but only the DNA molecule or genophore, there has been diversity from simple to complex structures. In the *euglenoid* forms, nuclear division involves duplication of chromatids and their separation into daughter nuclei. However, the absence of the centromere and spindle is characteristic of the group. High chromosome numbers have been reported in several genera, up to even 177 in *Peranema trichophora*. The chromosomes, though organized into discrete units with the capacity for duplication and separation, could not be resolved into specialized functional segments. The huge chromosome number recorded in such an otherwise primitive group is a special feature. A similar, slightly advanced mitotic system is found in the *dinoflagellates*. Here, chromosome number is also quite high and characterized by a clear cut nuclear membrane. Notwithstanding such features, dinoflagellates do not possess the basic protein in the nucleus, the presence of which is characteristic of eukaryotic systems. The chromosomal framework is composed of continuous DNA fibrils, appearing beaded at the ultrastructural level. The term *"mesokaryota"* is applied to Dinophyceae on the basis of its chromosome structure.

Presence of high chromosome numbers in such a primitive group with very little differentiation is an index of the redundancy of genes. Progressive chromosome evolution over the dinophycean system is manifested in the green algae. In the Conjugales, chromosomes are well defined, spindle structure is well organized, and separation of chromosomes to the two poles is well regulated. But the level of functional differentiation of chromosome segments is rather poor. Nucleolar organizers are present, but centromeric function appears to be dispersed throughout the chromosome length. However, the use of the term *"diffuse"* or *"polycentric"* has been suggested. It is difficult to resolve two adjacent centromeres in a polycentric system. There is, however, parallel movement of chromosomes along the spindle.

In Chlorophyceae, in general, the chromosome number is rather low ($n = 4$–10) as compared to other groups of algae. In Conjugales, despite the record of $n = 84$ in certain species of *Spirogyra*, the number is not very high. But in view of the diffuse centromere, the extent to which such high numbers reflect polyploidy or fragmentation is not clear. Chromosomes with a diffuse centromere have the unique advantage of survival of fragments. It is remarkable that such a group of algae, with highly organized photosynthetic system, reflecting a complex chloroplast structure with spiral bands, has maintained its evolutionary complexity within comparatively low chromosome numbers. Evidently, imperceptible gene mutations and structural changes have played very important roles in their evolution.

Siphonales, with its multiple nucleate system, maintains the individuality of its nuclei, with discrete chromosome groups. Such behavior is in strong contrast to higher plants, which show well-defined uninucleate cells in a complex tissue system. In angiosperms, tapetal cells are mostly multinucleate, but they often show nuclear fusion, resulting in very high chromosome numbers. Absence of any fusion in a multinucleate system, such as *Siphonales*, is thus in strong contrast to the frequent fusion observed in multinucleate cells of angiosperms, as in the tapetum and endosperm. Since the group as a whole is multinucleate, it may have an inbuilt genetic resistance to fusion of nuclei, which would result in high polyploid chromosome numbers within the multinucleate cells. Since the organism is undifferentiated, such resistance may give it an adaptive advantage against the rigors of selection.

Both in Phaeophyceae and Rhodophyceae, the body structure is highly differentiated as also the alternation of generation and sexuality. The characteristic feature in the brown algae is the presence of well defined sex chromosome with a large X in female as in *Saccorhiza polyscides* of Laminariales, where the chromosome number range being 16–62 is quite high. The small x is associated with maleness. Rhodophyceae, on the other hand, shows a range of chromosome number starting from $n = 3$ in *Porphyra* to $n = 72$ in *Polydes* (Rao, 1956) (vide Cole, 1990). Polyploidy and fusion of nuclei as well as endoreplication, leading to polyteny, are common. Telocentric chromosomes, and structural heterozygosity indicate that structural alterations of chromosomes have been effective in speciation. A multiple sex chromosome mechanism with xx_2-x/y_2 heterochromatic chromosomes also shows an advancement in sex chromosome differentiation.

Fungi

Fungal chromosome, on the whole, does not yield any definite clue to the evolution of mitosis or chromosome structure; on the other hand, it shows a high degree of complexity, often comparable to higher angiosperms. Of the three fungal species studied extensively, in *Saccharomyces cerevisiae*, both haploid and diploid numbers are stable; in *Aspergillus nidulans* the haploid is stable and the diploid is unstable; whereas in *Neurospora crassa*, only the haploid number is found. Diploids are extremely transient. Auto-polyploids are more frequent and tri- and tetraploids have also been found.

Extensive work on chromosomes of *Neurospora*, the ascomycete, shows a well-defined nucleolar organizer at the tip of chromosome arm, often comparable to higher plants and animals, and a complex chromosome structure. Nucleosomes occur in the form of octamers in the chromatin surrounded by a DNA supercoil of 200 bp. In species of *Saccharomyces*, the split gene nature has been fully established, together with intervening sequenices. In several other genera such as *Aspergillus* and *Neurospora*, a similar situation exists. A significant feature in the fungal chromosomes is the presence of low number of repeated sequences compared to that of higher organisms. In *Neurospora crassa*, nearly 90% of the sequences are unique, whereas in the polymycete *Achlya* sp., the repeats are slightly more in number. In this respect, they are relatively more advanced than the prokaryote, where the entire sequence is almost unique. Along with nucleosomes, the typical uninemic and multirepliconic structure of eukaryotic chromosome is also present in fungi. On the basis of DNA value, nearly 20,000–40,000 average genes have been estimated to be present in this group. However, the principal difference between nuclear systems of higher eukaryotes and fungi is the presence of a nuclear envelope in fungi, though the spindle pole bodies and initiation and termination loci of microtubules are also recorded in fungi.

The meiotic behavior of chromosomes is highly evolved as in higher eukaryotes, including even the presence of a *synaptonemal complex* between paired homologous chromosomes at pachytene. These data at the ultrastructural, molecular, and structural level of chromosomes indicate a high degree of specialization in this otherwise simple eukaryote. In the presence of a comparatively lower number of repeated sequences, the fungi are slightly more advanced than prokaryotes, but in the complexity of chromosome structure, they resemble the most complex eukaryotic system. The presence of a continuous nuclear membrane, on the other hand, is a feature unique to this group. Such complexity in lower eukaryotes such as fungi is noteworthy. It is true that even with such a complex genetic architecture, fungi have not possibly contributed to higher forms in evolution. They may represent a degenerate, or at least a blind lane, in the evolutionary system. The extent to which this high level of specialization in an otherwise simple thallophyte, with hardly any scope for differentiation, is responsible for a blind pathway in evolution is a matter of debate. Possibly, the saprophytic and parasitic adaptation of the group is related to this complexity in genetic architecture.

The meiotic behavior is normal, but an unusual behavior has been recorded in *Pyronema confluens*, a member of Saprolegniaceae. In this genus, during meiosis, heterotypic division is preceded by homotypic division and not vice versa.

Aberrations in meiotic behavior, as well as polyploidy, have been recorded in species of *Xylaria, Alomyces,* and *Cyathus.* Aneuploidy is recorded in species of *Aspergillus* and *Saccharomyces.*

Bryophytes

Bryophytes occupy a unique position in the phylogenetic system in view of their potential in the involvement in the ancestry of vascular cryptogams. The plant body, though well organized and differentiated into photosynthetic and conducting systems is but a gametophyte, containing haploid chromosome number. It is the only category of plants, where the chromosomes, though present in a single set, are in a well-differentiated phenotypic system, exposed to rigors of selection, characteristic of terrestrial environment. It is also the only group where the sporophyte containing the diploid chromosome number has attained a certain degree of complexity, but is dependent on the haploid system with single set of chromosomes. In spite of its uniqueness, the mechanics of chromosomal evolution in this group still remains ambiguous.

Bryophytes, as a whole, show comparatively low chromosome numbers, the lowest ($n = 4$) being recorded in *Takakia*.

The most prevalent chromosome number $n = 7$ is mainly restricted to *Bryopsida*, whereas in *Hepaticopsida* or *Anthocerotopsida* numbers are multiples of 9 and 5, respectively. In spite of the low number of *Takakia*, its relationship with the rest of the hepatics is still a matter of debate. In hepatics, the role of polyploidy is rather low, except in species of *Riccia*, where it is presumed to be associated with xerophytic adaptation.

In Bryophytes, in general, intra- and interspecific polyploidy have played a role in evolution. However, aneuploidy and reduction in number are claimed to be caused by structural changes involving principally unequal reciprocal translocations rather than loss of whole chromosomes, which may lead to lethality.

Sex chromosome

The plant body in this group being a gametophyte provides basic information regarding evolution of sex chromosomes, which were first reported in *Sphaerocarpus donnelii* and *Ceratodon purpureus*. In the former, the X chromosome was found to be the largest, and Y the smallest. However, later studies on dioecious species as a whole, in several bryophytes, have shown that the autosomal complement too plays a very important role in the expression of sex. This is proved by irradiation damage to X chromosome, which often results in maleness even in absence of Y. Evidently, autosomes, to a great extent, carry the maleness, though a male mobility gene has been claimed in the Y chromosome. In *Sphaerocarpus donnelii*, variable numbers of sex chromosomes have been recorded in diploid, triploid, and tetraploid sporophytes without affecting their normal behavior, indicating the importance of autosomes in the balance of sex determination. Taking all these factors into account, the term "sex associated chromatin" has been attributed to the so-called sex chromosomes of bryophytes (Smith, 1983).

A study of different genera of Marchantiales and Jungermanniales shows that chromosome evolution involved in monoecism to dioecism is associated with heterochromatization of small chromosomes in Marchantiales and large chromosomes in Jungermanniales and Sphaerocarpales. Sex-specific chromosomes are highly polymorphic at an intraspecific level, in different areas. The smallest bivalent in *Riccardia pirguis* is euchromatic and dimorphic, the larger one occurring in females, in populations from Britain. In Japan, they are entirely heterochromatic, and the metacentric chromosome occurs in males. In materials from North America and Central Europe, sex-specific chromosomes are absent. In Central Europe, on the other hand, races occur with no heterochromatin or with one or two heterochromatic chromosomes.

The extreme case is also on record, where sex-associated heterochromatin has been considered as a reflection of phenotypic response. Berrie (1974) reported that in the sporophyte of *Plagiochila praemorsa*, there are no morphological differences between two large chromosomes, but they differ in the degree of heteropycnosity in male and female gametophytes. According to him, the difference in amount of heterochromatin in the two sexes is a reflection of phenotypic response of chromosomes, microchromosomes, and accessories to male and female genomes (vide Smith, 1983).

Microchromosomes and accessories

A chromosomal feature of special significance is the presence of *microchromosomes* in several genera. There has been considerable debate as to their function and homology, but their regular separation in meiosis and their functional attributes differentiate them from *accessory chromosomes*. In general, they are almost half the size of the smallest chromosomes of the complement of mosses. Their association with sex determination has been claimed. The origin of these chromosomes, which contain a significant amount of heterochromatin, is a matter of debate. If, they are to be regarded as very small chromosomes, their origin from multiple sex chromosomes, is precluded. Such an origin holds good for accessory chromosomes.

The fundamental aspect of chromosomal evolution is manifested in *Pleurozium schreberi*, where centromeric activity has been recorded in more than one point. *Pleurozium* was regarded as representing an intermediate step between holocentric with diffuse centromeres and the monocentric states. The primitive centromere is visualized as associated with a simple chromosome block, undifferentiated by any constriction. The fusion of such blocks may lead to evolution of larger chromosomes, some with centromeric activities localized at the two ends, as in *Pleurozium*.

Such a fusion may lead to large chromosomes, with centromere at the telomeric ends. A concept of far-reaching implications is that fusion of two such telocentrics, may have led to the origin of metacentrics, otherwise termed as *Robertsonian fusion*, as demonstrated in some higher plant genera such as *Gibasis* of Commelinaceae. Robertsonian fusion visualizes origin of symmetry from asymmetry and metacentric from acrocentrics.

In general, structural changes of chromosomes in bryophytes are marked in different genera, including *Pellia, Riccardia,* and others. Despite the fact that trend of evolution is often difficult to determine in terms of evolutionary changes in chromosome structure, certain trends are distinct. For example, *Pellia rotandifolia,* with high degree of asymmetry in karyotype, has less heterochromatin. Evolutionary progress has been associated with reduction in heterochromatic segments. The saprophytic species *Cryptothallus mirabilis* has a very low heterochromatin content and is considered as advanced as compared to the allied species *Riccardia penguis.* The advance in evolutionary progress is associated with decrease in heterochromatin. The karyotype evolution in this group has thus involved loss of small heteropycnotic chromosomes and heterochromatin, along with changes from symmetry to asymmetry, through structural alterations.

Pteridophytes

Pteridophytes present a striking contrast to bryophytes, where every major group has a very high chromosome number. The high degree of polyploidy noted in several genera belonging to primitive groups indicates their relative antiquity. This phenomenon, coupled with the fact that very low chromosome numbers have been recorded in several genera, suggests a distant homology with flowering plants in terms of cytological behavior. Extensive hybridization and polyploidy have evi-

dently given rise to the modern fern genera. Even in primitive orders such as *Psilotales, Equisetales,* and *Ophioglossales,* chromosome numbers are unusually high. The highest number is in *Ophioglossales,* where the number is more than 1200 chromosomes. Regarding antiquity, ferns are characterized by a high chromosome number, accumulating in ancient forms. The case of *Ophioglossum* is an example where a high degree of polyploidy has been reached without lethality.

From the evolutionary point of view, ferns exhibit a marked adaptation in their chromosome behavior. Autopolyploidy is maintained in many cases, through bivalent formation. Evidently, the capacity of multivalent formation has been eliminated through gene mutation. In fern, diminution in chromosome size associated with polyploidy has been suggested to be related to physiological readjustment, enabling the organism to sustain successive polyploidization.

In pteridophytes, contradiction is often recorded in certain criteria of evolutionary progress, especially in species of *Psilotales, Lycopodiales, Equisetales,* and *Ophioglossales.* Radical reduction in chromosome number too has been induced through different cytological mechanisms, including fragmentation and translocation. There are several species with very high chromosome numbers in otherwise primitive groups, as mentioned above. A contrary situation is noted in genera such as *Selaginella* ($n = 9$), *Isoetes* ($n = 10$), *Hymenophyllum* ($n = 13$), and *Osmunda* ($n = 12$), where chromosome number is rather low. It is likely that in species where very high chromosome number has been recorded, the counterparts with low numbers are eliminated in evolution.

The relative roles of high and low chromosome number in evolution are reflected in pteridophyte evolution. The genus *Selaginella* with low number of $n = 9$ has almost 800 species, whereas *Equisetum* with $n = 108$ has only a few species. The evolutionary capacity of *Equisetum*, so manifested, may be due to its high load of polyploidy, which is the almost general rule in ferns. It is presumed that pteridophytes can hardly be regarded as a source of evolution of major flora on earth. The high chromosome number in ferns might have forestalled any adventure in evolution. The relative simplicity of most fern genera, with body structure much less differentiated than that of the phanerogams, is not commensurate with the number of genes expected to be present in such high numbers of chromosomes. A large number of genes that control differentiation are distributed throughout the genome in a high-numbered complement. However, the required number of genes may not necessarily be reflected in the size of the chromosome complement, as extra genetic material is as present, due to redundancy. In ferns, such redundancy does not remain restricted to the segmental level of chromosomes, but may involve entire chromosomes. The question is whether all such high chromosome numbers are necessary for existence and vital functioning of all fern species, where they are found.

Gymnosperms

Chromosomes of gymnosperms present certain unusual features in their evolution. In the living gymnosperms, several haploid numbers have been recorded in only two families, such as, 8, 9, 11, and 13 in Cycadaceae and 9 to 19 in Podocarpaceae.

The remaining families are characterized by single base numbers, such as 12 in Ginkgoaceae, Cephalotaxaceae, Taxaceae and Pinaceae; 13 in Auracariaceae; 10 in Sciadophytaceae; 11 in Taxodiaceae and Cupressaceae; 7 in Ephedraceae; 21 in Welwitschiaceae; and 22 in Gnetaceae. Structural alterations as well as polyploidy are suggested to be responsible for evolution. The changes involve segmental interchanges as well as para- and pericentric inversions, all principally leading to position effect.

The principal controversy centers around the direction followed by chromosomal changes. The cycads, though undoubtedly a primitive order, contain mostly acrocentric chromosomes, whereas conifers or podocarps have a large number of metacentrics. In the angiosperms, the situation is quite different, as most primitive families exhibit a symmetrical karyotype, and evolution has been assumed to run from symmetry to asymmetry, in general.

In view of the presence of acrocentrics in cycads, it is claimed that chromosome and morphological evolution have proceeded along two different directions in gymnosperms. However, if the suggestion that centric fusion, or Robertsonian change as the principal feature in evolutionary advance in gymnosperms, is accepted, the contradiction between morphological and chromosomal evolution is resolved. Both centric fission and fusion as well as fragmentation, have played very important roles in evolution of chromosomes, at least in certain families of gymnosperms. In cycads, evolution might have principally involved fission of chromosome ends. The direction of the change and its stabilization might have had adaptive implications in space and time.

Angiosperms

In the angiosperms, chromosome studies have been carried out extensively, both in dicotyledons and monocotyledons. The changes so far recorded involve both the structure and number of chromosomes. *Karyotype orthoselection* has been almost absent in the majority of flowering plants, except in certain families such as Agavaceae and Aloineae, where the typical karyotype with asymmetry and constant base numbers characterize the entire group. Otherwise, chromosomal alterations occur at all levels of taxonomic hierarchy, including families, genera, and species, and even at an intraspecific level. Moreover, B chromosomes and isochromosomes have arisen in certain genera through misdivision of centromere. Diffuse centromere, a primitive character, originally located in species of *Luzula* by Camara and associates has been located in other taxa, mainly those belonging to Juncaceae and Cyperaceae. However, in some groups, localized centromere has been noted in these families. It is thus difficult to outline the specific chromosomal changes that have occurred during the evolution of angiosperms, even though they have aided in solving problems of taxonomic dispute in many cases. In general, however, polyploidy, aneuploidy, and structural changes of chromosomes have been operative in the evolution of different taxa. The tendency of evolution has been from symmetry to asymmetry, accompanied by diminution in chromosome size. Chromosome changes induced through fragmentation, translocation, inversion, associated with interspecific hybridization as well as polyploidy often lead to numerical alterations as well. The importance of structural alteration of chromosomes in the evolution of higher plants is now being gradually appreciated with advances of methodology for the study of chromosome structure.

Diffuse centromere

In species with a diffuse centromere, distinct evidence of fragmentation of chromosomes, leading to increase in chromosome number and speciation, has been claimed. The survival of so-called acentric fragments on the spindle, their role as distinct chromosome in species of *Luzula* with high chromosome number, the total amount of DNA remaining same as with diploid and polyploid, and pairing of long chromosome with multiple short chromosomes in the hybrid distinctly indicate chromosome fragmentation as a basis of speciation.

Repetitive DNA

In the chromosome structure, origin, increase and decrease of repetitive sequences have also been the principal features of evolution in higher organisms. Various epithets have been applied to these sequences, including junk, non-essential, and selfish. But most of the repetitive sequences have in common properties of amplification, dispersion, and mobility. As such, the term "Dynamic DNA" was proposed. It exerts a dynamic control over all nucleotypic functions, including cell division and chromosomal integrity. Multiple copies have a regulatory role and provide scope for generation of genetic variability, whereas mobility and dispersion are strategies to withstand total obliteration during evolution. Such sequences are essential for influencing genes required for physiological adaptation. Thus, evolution of sequence complexity of chromosomal DNA with special reference to repeats and the dynamic influence had been the significant features in chromosome evolution in eukaryota, with special reference to higher plants (Sharma, 1978, 1983).

Synteny

A new focus has now come up in the concept of chromosome evolution in the plant kingdom, with the recent discovery of colinearity of gene sequences in chromosomes of widely different plant genomes. Studies carried out in recent years, especially in the cereals belonging to the grass family, revealed a remarkable conservation of gene order. The data on wheat, rice, and maize indicate that excepting for major translocation, the gene order is essentially similar in these species of grasses despite gross differences in the genome size.

On the basis of a comparison of organization of genes of different cereals, it has been proposed that it is possible to generalize genome structure of the ancestral grass which gave rise to wheat, barley, maize, sorghum, and rice nearly 60 million years back. Linkage segments forming blocks were found to be 19 in rice, which can be used to interpret all the species of cereals, including maize and sugarcane. This colinearity and conservation of gene order is otherwise termed as *synteny* (Moore, 1995). It is presumed that ancient grass might have had just one chromosome pair. Assuming breakage in different sites, inversion and amplification, genomes of maize, wheat, barley, sorghum, and millets can be constructed on the basis of these common linkage blocks, using 19 rice linkage segments. The maize genome with $2n = 20$ chromosomes is supposed to be a tetraploid rather than a diploid. The processes that have played a very important role in evolution from the ancestral genome are duplication, translocation, inversion, fragmentation, and specially amplification of repeated DNA sequences of chromosomes.

The concept of synteny suggests that 10,000 species of grasses have evolved from a set of gene blocks located initially and possibly in one chromosome. Evidences are forthcoming of syntenic arrangement of genes in chromosomes and conservation of gene order of different groups of plants and animals. In this concept, the differences in chromosome size and increase in DNA amount have, to a great extent, been attributed to repeated DNA sequences. The analysis of syntenic arrangement enables prediction of similarity in related genera. This understanding of gene arrangement of chromosomes helps in delineating affinity and status in phylogeny as well as facilitating the gene transfer experiment.

It is known that there is great variation in size and structure of the genome in flowering plants, which ranges in size from ~ 38 Mb in *Cardamine amara* (a crucifer) to > 87,000 Mb in *Fritillaria assyriaca*. Despite this large variation in gene content, the gene order, and even the gene function, are fairly conserved in a wide range of plant species. Detailed molecular maps are now available in all major crops, which allow comparative genetic analyses in several groups of plants, involving families such as Poaceae, Solanaceae, and Brassicaceae. This can be judged using the examples of the grass family and that of tomato and pepper from Solanaceae. It has been shown that all genomes of grass species can now be described in terms of their relationships to a single reference genome, i.e. rice. Less than 30 linkage blocks are needed to represent genomes of all grass species, including wheat, rye, barley, oats, maize, sorghum, sugarcane, rice, wild rice (*Zizania palustris*), foxtail millet, pearl millet and finger millet (*Eleusine coracana*). Similar comparative genomic analysis has also been under-taken in the family Solanaceae, where the maps of tomato, pepper, and potato were compared, suggesting that 18 homologous linkage blocks cover 98.1% of the tomato genome and 95.0% of the pepper genome. An ancestral genome was also proposed, which through translocations, inversions, dissociations, or associations of genomic regions differentiated into tomato, pepper, and potato genomes (vide Gupta, 2001).

Just as DNA sequence studies have helped in demonstrating synteny and colinearity of the genome, another aspect of chromosome evolution has been aided by sequence analysis. Bennett et al. (1995) analyzed the organization of two highly repeated ribosomal DNA sequences in *Zingeria biebersteiniana* ($2n = 4$). The studies show three 18S–28S rDNA sites in short arm of chromosomes. These sequences normally do not occur more than at two sites in a chromosome. As such, it is presumed that these three sites in a chromosome have originated through tandem fusion, which would suggest that the ancestral genome might have contained more than two chromosomes. In other words, the origin of the lower number might be traced to fusion of higher number of chromosomes in the ancestral genome.

CHAPTER 6

Chromosome Behavior in Differentiation

The influence of chromosome on the regulation and control of differentiation of organs, is an expression of its dynamic behavior. From this aspect, even the sudden switch over from mitosis to meiosis in the germinal line, and further reversion to mitotic behavior of the gametophyte nucleus, are remarkable examples of dynamism. This dynamic behavior of chromosomes permits recombination of characters, at the same time maintaining the constancy of chromosome number in the hereditary cycle. After the cessation of meristematic growth, represented by active cell division, differentiation of the nucleus is initiated. With the onset of maturity, the tissue enters into the adult stage and does not show active mitosis. But gene action, necessary for enzyme synthesis, is at the optimum level at this stage as enzymes control all steps in differentiation. Because of the comparative plasticity of plant tissues, which often quickly respond to external influences, the mechanism of differentiation and its control could be well worked out (Nagl, 1976a).

By inducing division in the differentiated tissue through different chemicals, the polytenic state of differentiated nuclei, as in anther tapetal cells and suspensor (Guerra et al., 1996), has been demonstrated, indicating their origin through endomitotic replication. Huskins and Steinitz (1948) induced division in energic nuclei through indolyl acetic acid treatment and recorded their polyploid constitution. Similar induction of division was also obtained in differentiated nuclei (Sharma and Mookerjea, 1954) of *Allium cepa* roots through sugar treatment. Deficiency of a sugar moiety of nucleic acid was suggested to be responsible for polytenic state, and separation of strands was achieved by addition of the sugar. The role of chromosomal histones in spontaneous transition from mitotic to endomitotic state has also been demonstrated (Nagl, 1975) and the experimental induction has been secured through actinomycin D and histones. Specific sat-DNA has been identified, originating through DNA amplification in the polytene chromosome suspensor. Differential replication of the standard satellite DNA in these organs has also been claimed (cf. Ingle and Timmis, 1975). Division has been induced in adult nuclei by 2,4-dichlorophenoxyacetate (Sen, 1974b), in addition to IAA, but such nuclei in differentiated region exist both in diploid and polytenic states, and on induction of division, both diploid and polyploid metaphases are noted.

Variability in Relation to DNA

Cook (1973) noted that cellular differentiation is controlled by the development of the higher order of polymorphism of chromosomes. Variability of the DNA content from organ to organ has been recorded in various species. The existence of metabolic DNA in addition to genic DNA, playing a significant role in differentiation, has been claimed. In *Petunia hybrida*, metabolic DNA synthesis has been recorded in the beginning of the G1 phase (Essad et al., 1975). Similarly, transient DNA satellite has been observed in the differentiating pith tissue (Parenti et al., 1973). It has also been suggested (Pelc, 1972) that different forms of organization can be regarded as modifications of systems in which metabolically active DNA plays an important role. In growing roots of *Vicia faba*, DNA content was three times greater in the elongated region as compared to that of meristematic tissue. Differential DNA replication, under or extra, is regarded as a part of genome activity, especially in the heterochromatin of several plant species. Variability in the DNA content in different organs is shown through microphotometric analysis. For example, in *Vicia faba*, *Pisum sativum*, and *Nigella sativa*, chromosomes of different organs show clearly different DNA values,

the shoot cells showing greater amount than that present in the roots (Sharma, 1976, 1983). This quantitative variation occurring in tissue metabolism of content of DNA as compared to that of meristematic zone has been confirmed through DNA isolation. Such variations are due to differential amplification termed as "ontogenetic increase" by Nagl (1976b).

Variability in Relation to Protein

Of the components that enter into composition of chromosomes, in several organisms, lysine rich histones are often replaced by arginine rich histones. Quite sizeable portion of sperm protein has also been shown to be non-basic (Prescott, 1970). Ruderman et al. (1974) noted that the amount of histones, that associate with DNA of chromatin during embryogenesis, differ characteristically from stage to stage. The highest histone/DNA ratio, during endoreduplication, in metaxylem cell line in *Allium cepa*, is also on record (Innocenti, 1975).

Necessity of Endomitotic Replication

A large number of species, both plants and animals, undergo endomitotic replication during differentiation (Nagl, 1976c). Several types of plant tissues, including suspensor and endosperm, show such behavior. It is claimed that endomitosis is needed for high rate of synthetic activity (Millard and Spencer, 1974). Evans and Van't Hof (1975) recorded polyploidy in certain differentiated organs but not in others, as in species of *Pisum*, *Triticum*, and *Helianthus*. All these data (Nagl, 1976b) suggest that increase or differential amplification of basic DNA is often needed for differentiation and specialized function of certain cells in eukaryotes. It may be achieved through repetitive DNA, generative polyploidy, or even linear doubling of chromosome size through telomere replication, resulting in a single long palindrome or inverted repeats. Increase may be needed, which is often necessary for specialization and adaptation. However, it is not obligatory, as the entire process of differentiation is under genetic control. A negative relationship between C value and degree of polytenization has been claimed. If the amount of DNA is low without any variability, DNA endoreplication or polyteny is an evolutionary strategy, compensating for the lack of phylogenetic increase in nuclear DNA (Nagl, 1976b). Endomitosis, leading to fresh DNA strands, allows uninterrupted RNA synthesis without being discontinued through chromosomal coiling as in mitosis. However, it has been shown through uridine uptake that only a single strand of DNA is active for transcription (Sen, 1978) during differentiation. It is suggested that the limited transcribing life of the DNA molecule or the need of extensive transcription may explain the occurrence of polyteny and polyploidy, which may provide with the supply of fresh strands for transcription. Therefore, endomitosis during differentiation is an example of the dynamic property of the chromosomes (Sharma, 1975a, 1976). It is implied that species having phylogenetically increased nuclear DNA through polyploidy or high degree of repetitive DNA may not resort to endomitotic replication (photographs 6.1–6.2).

Absence of polyteny in certain organs and its presence in other organs of the same individual need not be considered as a negative evidence for endomitosis. The

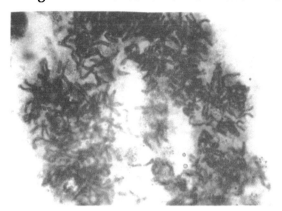

Photo 6.1 Very high chromosome number following endoreplication in endosperm of *Nothoscordum fragrans*.

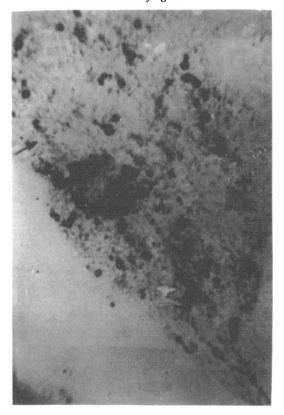

Photo 6.2 Large polytenic cell in suspensor of *Psophocarpus tetragonolobus*.

transcribing capacity of different genes, and as such chromosome segments, is also under genetic control. There is no evidence to show that all the DNA cistrons of a genome should have the same transcribing capacity. In that case, endoreplication would be necessary in cases where the transcribing capacity is such that it cannot cover the entire differentiating period of the tissue concerned. Genetic control can explain the absence and occurrence of polyteny in different organs.

In the chromosomal control of differentiation, discovery of the use of antisense DNA (Izant and Weincraub, 1984) has opened up the possibility of utilizing such sequences to block plant differentiation at certain phases of growth.

The technique is based on the principle of blocking the informational flow of DNA to protein, by introducing one RNA strand, complementary to the sequence of the target RNA. On the basis of standard base pairing, a RNA duplex is formed between messenger and antisense RNA and translation is blocked. Several reports have been published of using antisense RNA in the plant system to check certain paths of differentiation. In fact, to increase the shelf life of fruit, antisense RNA has been used to check the activity of genes responsible for ripening, which is associated with ethylene production. Because of the totipotency of plant cells, an entire plant can be originated from a single cell with an antisense gene construct (Sandler et al., 1988; Van der Krol et al., 1988).

The discovery of the effect of antisense RNA led to search for natural antisense RNA in plants, as noted in prokaryotic system and animals. In plant system, antisense RNA of amylase has been found in barley, which covers the full length of both type A and type B of amylase in RNAs. The hydrolytic enzyme digests the storage content of endosperm for use by the seedling during differentiation (Rogers, 1988). Since then, several antisense RNAs have been obtained in cell cultures. In nature too, antisense RNAs play a very important role in differentiation. It has been suggested that at the chromosomal level, antisense RNAs are transcribed in loci different from that of RNAs, as both sense and antisense amylase transcripts are present in equal amounts in the aleurone tissue of barley.

The discovery of natural antisense RNA in plants may indicate its critical role in differentiation. Differentiation in the plant system is, to a great extent, controlled by repression and derepression of genes at the chromosomal level. As such, it is likely that repression of the expression of certain genes and enzyme products, may be facilitated by an antisense messenger produced from a chromosomal locus at certain phases of development and differentiation.

Differentiation to Dedifferentiation

At the chromosomal level, replication provides the basic DNA strand needed for continued transcription. It is necessary since transcription, translation, followed by enzyme synthesis, ultimately control differentiation.

However, a differentiated cell can also undergo dedifferentiation, forming an unorganized mass as in the case of in callus. McClintock in her Nobel Address in 1983 stated that formation of dedifferentiated callus in culture is the manifestation of stress response caused by isolation of cells in culture. As such, in vitro growth is to be regarded as a phenomenon involving stress response. This implies that a differentiated tissue can be dedifferentiated through induction of stress. If the stress is very strong, it often leads to somatic instability or chromosomal variability. Lately, it has been found that along with somatic instability or chromosomal variability, there may be regulated methylation of cytosine, which in turn can hamper replication at the chromosomal level and may lead to mutation.

Genetic Control of Differentiation– Chemical Basis

As all cells of an individual contain similar DNA sequences, the differentiation is expressed, as discussed above, in terms of repression and derepression of gene sequences in different organs. It is recognized that histone and nonhistone sequences are associated with repression and derepression.

Lately, there are new evidences to explain the mechanics of differentiation at the cellular level. Repression of genes can take place at the DNA level or at the RNA level through RNA editing or alternate splicing, and even at the protein level. At the DNA level, it may be through DNA methylation, which is now considered to be an associated phenomenon in differentiation. In rice, adenine methylation has been recorded, which varies from organ to organ. As such, methylation, standing against transcription, occurs at different loci or chromosomes in different organs, leading to differential organogenesis. In addition, the role of transposable sequences in methylation during differentiation is not very clear. However, methylation has been shown to be a barrier against transposon-induced mutation, responsible for failure of replication. Cytosine methylation in somatic instability is triggered by transposons. McClintock too claimed that transposon or mobile sequences in chromosomes often play an important role in programmed cell differentiation. However, transposon movement is generally induced through stress effect (Federoff, 1984). The extent to which differentiation may involve some programmed intracellular stress on transposons in chromosomes, is yet to be studied.

RNA Editing and Alternate Splicing

In a number of species, including species of *Oenothera*, it has been recorded that transcribed RNA undergoes significant sequence changes during processing of the mitochondrial and chloroplastid transcripts after transcription. Similarly, the process of splicing, which involves excision of intron (the nonessential sequence) – and joining of the exon to the essential one may not always follow identical exon joining in all cases. As such, splicing may be *alternate* (Herbert and Rich, 1999). Such alternate splicing and RNA editing may generate different RNAs from a single gene sequence. The outcome of a single RNA processing event may regulate the outcome of another, which ultimately affects the expression and composition in differentiation (Gesteland and Atkins, 1993). In these cases, the messenger RNA that is translated in not identical with the messenger RNA that was originally transcribed. This process is known as *RNA editing*. It is now considered that such RNA editing may play an important role in differentiation (Benne, 1996) and may differ from organ to organ. Editing involves, in general, change of specific nucleotide sequence, especially cytosine and uracil, controlled by *"guide RNA."* This method of control is different from antisense approach. In the latter, in contrast to editing, the sequence remains the same and only the action is blocked.

Chromosomal control as expressed through the chromosome protein level is now being studied through the fluorescent antibody technique. It is well established that a fraction of proteins, histones, and nonhistones associated with chromosomes often undergo changes during differentiation, e.g. protamine in sperm and histone in body cells. Evidently, such changes are involved in the control of differentiation. The antibody against a specific protein, if tagged with a fluorescent compound, permits the detection of the presence or absence of the target protein in chromosomes at different phases of growth. Such studies have been carried out extensively with the centromeric protein.

In chromosomal control in differentiation, both structural and behavioral characteristics of chromosomes are involved. Endoreplication meets the need of fresh DNA strands, which are necessary for continued transcription and translation. Antisense and DNA methylation help to control gene expression underlying differentiation. Both histone and non-histone proteins of chromosomes undergo modification during development and differentiation, as detected through immunofluorescence antibody studies. Heterochromatin condensation is also an important factor in the control of differentiation. Thus, the process of organogenesis is controlled at the level of chromosomes through their chemical component as well as differential behavior.

CHAPTER 7

Nuclear DNA and Plant Evolution

The role of the basic genetic material – the DNA – in evolution is well established (Flavell, 1980; Price, 1976). Each genome is characterized by a constant DNA amount. Swift (1950) introduced the term "C value" (C standing for constant) to refer to the entire haploid complement of DNA in the nucleus. Thus, the nuclei of cells arising out of meiosis and containing unreplicated DNA contain the 1C DNA value, while the nuclei of cells containing fully replicated copies of two parents each, such as those at pachytene of meiosis, contain 4C DNA amount.

On the basis of the fact that DNA is the genetic material, the C value should show an increase corresponding to progressive complexity and differentiation in higher organisms. In nature, however, such a correlation does not necessarily exist. Increase in complexity is not necessarily associated with increase in DNA amount. The absence of correlation between amount of DNA and evolutionary complexity is otherwise known as "C value paradox." For instance, rice and rye, belonging to that highly evolved family Gramineae, have low amount of DNA (0.6 and 9.6 pg, respectively), as compared to less evolved lilies, which have 200–300 pg as the DNA amount.

DNA Value and Polyploidy

Since the establishment of the DNA nature of genes, it is accepted that increase in gene dosage should be associated with a corresponding increase in the DNA value. In other words, there will be multiplication of DNA value from haploidy to polyploidy. Such a logarithmic increase has been recorded in several genera. However, several species of plants do not show such a corresponding increase. On the other hand, a regular diminution in chromosome size has been recorded in several genera of flowering plants. The correlation of DNA value with polyploidy can best be analyzed if such duplication involves a single genotype or species. Such a situation is possible where a species has a series of polyploid and aneuploid cytotypes, as in grasses. But mostly analysis of such a correlation has been made depending on interspecific variations such as polyploidy in different species within a genus as in

Chlorophytum. These data do not give a true picture of correlation, as genotypic difference may influence the DNA value as well. At an intraspecific level, where the same genome is involved, such a correlation, if any, becomes meaningful and clearly reflects an increase or decrease in DNA value with difference in gene dosage.

Range of DNA Value and Function

Comparison of result of various plant species, starting from algae to angiosperm, reveals striking variation in genome size between species. The C value shows more than a 600-fold difference within the angiosperms alone, ranging from 0.2 pg in *Arabidopsis thaliana* to 127 pg in *Fritillaria assavriaca* (Bennett and Smith, 1991). Intra- and interspecific genome sizes have also been analyzed in several cultivated taxa (Cavallini and Natali, 1991; Greilhuber and Ebert, 1994). In most cases of such variation, the difference may be attributed to the repeated sequences of DNA, which exist in large amounts. Only a fraction of the total amount of DNA codes for structural proteins, whereas the rest are repeats – non-coding or coding for non-specific effects. Noncoding sequences are present in heavy amount in several families of flowering plants and constitute a portion of the total DNA amount. A similar behavior is also responsible for often heavy difference of DNA, at even the intra- and interspecific level. In several plant species, including *Secale cereale* DNA is repetitive in nature (Flavell, 1980).

The large amount of DNA, which is highly conserved, may also contribute to genotypic difference between species and varieties. Despite their almost universal presence, these sequences have been termed as *junk, trivial,* or even *selfish.* Their presence, it is assumed, is ensured by their property of reproduction. According to Doolittle, selfish DNA is a piece of DNA that is selected only for intragenomic survival.

To differentiate between specific and nonspecific effects of the amount of DNA in the life cycle of plants, the terms *"genotypic"* and *"nucleotypic"* DNAs have often been used. While the former involves structural genes coding for specific and mostly qualitative characters, the nucleotypic DNA, is generally associated with a number of biophysical parameters of the cell. Such parameters may include nuclear volume, chromosome volume, generation time, padding to keep chromosome in the folded state, and such other characters (Egel, 1981; Price, 1976; Sharma, 1983).

Variation in C value of DNA at inter- and intraspecific level is often correlated with diverse characters at the level of chromosomal, nuclear, cellular, tissue, and even organisms (Bennett and Leitch, 1995; Greilhuber, 1998). Variations of DNA amount in chromosomes between different species is often associated with differences in length and volume of chromosome, period of cell cycle, duration of meiosis, radiosensitivity, and other characters (vide Bennett and Leitch, 1998).

DNA Value, Chromosome Size, and Heterochromatin

Differences in chromosome size and DNA value, from diploid to polyploid, have been noted in several taxa, including species of *Vicia, Commelina, Pisum, Lens, Lathyrus, Ophiopogon,* and *Chlorophyton.* It has been also recorded in several colchitetraploids, such as in species of *Nigella, Callisia* and *Vicia*. The sudden decrease in size, even in the C_1 generation of the colchiploids, indicates the efficiency of genotypic control exhibited in quick response to the changed setup (Sharma, 1976). However, such overall diminution in size, at least in some genera, may or may not be associated with decrease in DNA content. Darlington (1965) suggested that reduction in chromosome size, which is genotypically controlled, is an adaptation to a decrease in size of the cell or to an increase in the number of chromosomes. In cases where there is no decrease in DNA, reduction in size can be attributed to heavier condensation, possibly of heterochromatic segments. The increase in gene dosage is expected to meet the need for adaptation to extreme stress situation. In the changed setup in polyploidy, heterochromatin representing mostly repeated DNA sequences, which otherwise confer adaptability, may prove to be redundant and becomes highly condensed and inactive. On the other hand, there are cases where size reduction is also associated with decrease in DNA content. Such a decrease, as noted in species of *Aglaonema* and related genera in Araceae, has been attributed to elimination of heterochromatin or rather repeat DNA sequences. The elimination of accessories – the cytological embodiment of repeated sequences – consequent to polyploidy, as recorded in species of *Urginea* and *Allium* confirms the above contention. The plants of *Allium stracheyii* of the temperate region, when transferred to plains, gradually attain polyploidy and lose their B chromosomes (Sharma and Aiyangar 1961).

Range of DNA Value in Different Groups

The values of nuclear DNA often vary widely among different groups. In general, tree species exhibit low DNA value in chromosomes. The situation is contrary to the values noted in gymnosperms, where high amount in trees is rather common. Comparatively less selection pressure in the temperate gymnosperms, and long generation time, have been considered as possible factors responsible for large chromosomes in temperate species. The high DNA value is also a reflection of ecological status. Tree species with low DNA value principally belong to semiarid regions where stress conditions are prevalent.

There is, in general, an increase in DNA content with increase in chromosome number. About 15-fold increase is often associated with a 17-fold increase in chromosome number in some legumes. Polyploid species often show a proportionately low amount in several genera, including *Acacia, Luzula, Typha,* and *Cenchrus*. In polyploid Acacias, there has been a visible reduction in the DNA content per genome as compared to that of diploids. The values in polyploids in *Acacia* are rather lower than the multiplied value expected in direct proportion to diploids. Such low value and comparatively shorter chromosome in polyploids have been reported in several genera. There has been a consistent decrease in chromosome size vis-a-vis DNA value, with polyploidy, in different species of plants.

Polyploidy itself, due to duplicate gene dosage, confers adaptive advantage to species against extreme climatic conditions.

Simultaneously, stress situation requires low generation time to complete the cell cycle, which can be facilitated by short size of chromosomes. Stresses due to temperature, water, and salinity often lead to adaptation of species with low DNA value; such low value as such is correlated with fast growth of the species.

The distribution of DNA in a species is not to be regarded as occurring at random. To a great extent, association with habit and ecological adaptation has been noted. The DNA of temperate species is generally more than that of the tropical representatives. In gymnosperms of the temperate regions, as recorded by Lima-di-Faria (1986), the amount is rather high, such as

Taxus baccata:	22.3 pg
Juniperus chinensis:	30 pg
Picea albertina:	85 pg
P. abies:	38 pg
P. sitchensis:	38 pg
Larix decidua:	30 pg
Pseudotsuga douglasii:	25 pg
Pinus sylvestris:	30 pg

Similarly, temperate herbaceous perennials show very high DNA value:

Hyacinthus orientalis:	94 pg
Tulbaghia violacea:	85 pg
Trillium grandiflorum:	92 pg
Tradescantia virginiana:	62 pg

In cereals, in general, the DNA content is comparatively low, as recorded by several authors. Despite low DNA content, the value of repetitive DNA content is quite high in cereals, being almost 50% in rice, 68% in maize, 75% in rye, and 50% in Triticale genomes. In cereals, though the value of DNA is low, yet in monocotyledons as a whole, the range of haploid DNA

value is 1–200 pg in annual forms. Further, it has been noted that plant species with less than 3 pg of DNA do not generally have satellite or high homogenous repeats. This is specially manifested in heavy amount of satellite DNA in dicots as compared to its very rare presence in monocots.

DNA value, plant habit, and intraspecific differences

Besides terrestrial species, a moderate DNA value has been recorded in a few aquatic taxa such as *Najas*, *Eichhornia*, and *Monochoria*. These species with a moderate DNA amount belong to the primitive Helobial families of monocotyledons, namely Najadaceae and Pontederiaceae. Plants grow in nutrient-rich aquatic environment. In aquatic herbs, on the other hand, a different trend is noticed, showing decrease in DNA with increasing altitude. A noteworthy feature of the DNA value of aquatic species is their low content in individuals growing at higher altitudes. In species such as *Sagittaria sagittifolia*, the distribution of which is rather wide, ranging from the plains to subalpine regions, the DNA value in high-altitude biotypes is rather low with comparatively low chromosome number as compared to those in the plains. The constant occurrence of such specialized genotypes may have some selective advantage.

Most of the aquatic and semiaquatic plants, in general, possess a high DNA value. The aquatic environment presumably provides with less selection pressure, in a setup with adequate nutrient supply. The xerophytic *Euphorbia* species has a high DNA value, principally due to heavy polyploidy. The increase in gene dosage caused by polyploidy enables adaptation to extreme climatic stress. For the same reason, increased DNA, but with shortened chromosome size, has also been observed in several genera occurring at high altitudes. In general, plants inhabiting areas of high altitude, or rather, the alpine zone, mostly have short generation time, which is facilitated by the low content of DNA and a rapid mitotic cycle.

Heavy differences in DNA value at the interspecific level are principally due to polyploid and aneuploid cytotypes, as well as structural variants. A wide variation occurs in species of *Vallisnaria* and *Hydrilla*. It is attributed to abundant nutrition, favoring a broad spectrum of cytotypes against less rigors of selection in aquatic environment.

Polyploids with moderate to high DNA content occur mostly in several terrestrial monocotyledonous species as of *Cenchrus*, *Billbergia*, *Dioscorea*, and *Pandanus*. The moderate DNA amount, even in polyploid species, is itself an index of adaptation. The value in these species is possibly maintained at a moderate level under a selection pressure to cope with the cell size without, in any way, affecting the qualitative characters.

Interspecific differences in DNA have been recorded in several genera with different ecological preferences. A very high interspecific difference showing more than a six-fold increase occurs in *Euphorbia*, which is associated with a 10-fold increase in chromosome number. The wide difference in chromosome number noted in Euphorbiaceae is also correlated, to a great extent, with the vast range of habit that the family represents. Its representation ranges from dendroid to herbaceous types

as also the preference for moderate to extreme climatic conditions. The difference in chromosome number especially due to polyploidy is associated with increase in DNA content, though not necessarily in direct proportion.

DNA amount, cell cycle, and development

There are certain parameters of growth and metabolism with which DNA content is associated. High DNA value is related to slow development. Thus, annual species, in general, have a lower DNA content than perennials. This fact may be related to the absence of annual habit in gymnosperms. The nuclear DNA content of plant has direct relation with generation time, which affects the rate of plant development, i.e. period of cell cycle, meiotic duration, pollen maturation time, and such other features. It is also generalized that species with medium or high DNA value are not ephemerals and those with high values are, in general, obligate perennials. However, the species of *Dioscorea*, though perennials, perennate through bulbs and behave as annuals. Selection pressure for completing life cycle at a short period as such is obligatory.

In extreme climatic situation of alpine Himalayas, flowering plants with high DNA values are almost absent. Moreover, it is difficult for species with heavy amount of DNA to establish from seeds in the short growing season in extreme climatic conditions. In high altitudes, selection of species with low DNA value as such is generally favored in view of the requirement of low generation time in such an environment.

Since high amount of chromosomal DNA and slow mitotic cycle evidently do not permit fast growth, these two parameters are often used for screening of genotypes capable of fast growth which is an essential requirement for energy plants. Extensive studies carried out on species of *Acacia* show that large amount of DNA is related to slow mitotic cycle and slow growth. Similarly plants growing under stress conditions of arid and semiarid regions do not have a high DNA content even after attaining polyploidy for increased tolerance (Mukherjee and Sharma, 1993). As far as evolutionary rate is concerned, highly evolved species with low DNA content may represent a blind lane in evolution as compared to less advanced species with more DNA content (Price, 1976).

Parameters of Genetic Diversity

In DNA, the base pair composition can be used to evaluate the taxonomic and phylogenetic characteristics of different taxa. To study the base composition in terms of GC content, thermal denaturation of DNA extract is carried out for the determination of T_m, the melting temperature. This value can be used to calculate the percentage of molar GC content.

Lower organisms show much more variability in base composition of DNA and higher GC content than the more advanced forms. The species relationship and phylogenetic connection can be traced even at the genotypic level through GC content of DNA. Therefore, this variation may serve as a good parameter for analysis of genotypic diversity.

The GC content of DNA is extremely variable in obligate perennial as compared to that of facultative perennials and annual species. Obligate perennials are also characterized by high GC content and often associated with specialized features such as woody habit. Facultative perennials occupy intermediate position in terms of GC content between obligate perennials and annuals as noted in species of Araceae. In this family, high GC content is considered as an adaptive advantage for survival in specialized habits.

Mechanics of Change in Evolution during Development

The mechanics of change in DNA content is manifold in the phylogeny of a species. They include, as discussed, differential polyteny, DNA doubling, gradual evolutionary accumulation of small deletions or duplications, regional disturbances in DNA replication, as well as increase or decrease of repeated sequences (Price, 1976). The repeats may originate out of saltatory replication, in tandem clusters, unequal crossing-over, as well as insertion or transposition. All these methods are the mechanisms through which DNA content undergoes alterations and accelerates evolution of species.

In general, polyploidy and repetitive sequences are considered to be mechanisms of phylogenetic increase of nuclear DNA. Such additional genetic elements can meet to a great extent the need of adaptations. However, such additional elements, if present in the genotypes of the species, are often adequate to meet the need for control of differentiation, i.e. genetic control of organ-to-organ diversification. In species where there is no phylogenetic increase either through polyploidy or high amount of repeats, the organisms develop an evolutionary strategy of ontogenetic increase in nuclear DNA through endomitotic replication. The need for amplification of genetic material for control of differentiation is well established, which accounts for ontogenetic increase or, more precisely, increase during development (Nagl, 1976a).

The process of histological and biochemical complexity is obviously associated with quantitative acceleration of enzyme activity. Both these aspects of metabolism require continuous transcription and translation. Regulatory sequences may play a significant role in repression and derepression. To meet the demands arising out of progression of complexity, additional amount of DNA may be needed. This extra DNA, whether for nonspecific or for specific function, is obtained through an increase in DNA amount in evolution by high degree of amplification or polyploidy (Sharma, 1985).

In plants higher in the evolutionary scale, a balance between phylogenetic and ontogenetic increase in DNA has been one of the principal features in evolution. The C value paradox can only be understood through an analysis of the basic features in evolution that underlie a phylogenetic increment up to a certain level, followed by genetically controlled amplification of DNA during ontogeny through endoreplication.

With the discovery of repeated sequences in DNA, it was realized that the variation of C value in angiosperms involves mainly an increase in the amount of repeated DNA sequences, which are again mostly nongenic and nontranscribing. The significance of C value is being gradually realized. Nongenic DNA probably accounts for 90% and perhaps even more of the DNA in the angiosperms with largest genomes. Genotypic difference

in C value of DNA is often correlated very closely with several phenotypic characters at nuclear, cellular, tissue, and even organismic levels.

Major interspecific variations in DNA amount in angiosperms have been correlated with the total volume of centromeres per nucleus, chromosome length or volume, nuclear volume, and mass and the volume of mature pollen grains. Correlations have been shown between DNA and number of chloroplasts per guard cell and the number of copies of chloroplast genome per mesophyll cell. Seed mass, rate and duration of DNA synthesis, minimum duration of the mitotic and meiotic cycle, and minimum generation time also show correlation. Even radiosensitivity and rates of induced mutation, ecological and phenological factors, and finally geographical distribution are also influenced. The phenotype is influenced by the genic content, and the physical effects of its mass and volume. The term "nucleotype" was coined to define those conditions of the nuclear DNA that affect the phenotype independently of the coded information. The large-scale variation in DNA amount and consequent effects show that C value is of fundamental significance for genotypes.

The C value is widely used in studies on cytotaxonomy and evolution. Considerable intraspecific variation in DNA amount has also been noted despite constancy in chromosome number and ploidy level. Such intraspecific variations are presumed to have adaptive significance through their nucleotypic effects on phenotypic and phenological characters.

Intraspecific variation in DNA amount is effected through two different ways in the karyotype. In certain angiosperms, such variation involves either the addition or deletion of certain absolute amount of DNA to each chromosome type, whereas in others, it may involve the addition or deletion of a constant proportional amount of DNA while maintaining the relative shapes of all chromosome arms in the karyotype constant (Brandham, 1983).

The study of C value of DNA is essential for studies on molecular biology as it provides an idea of the amount of repeats to be dealt with and choice of species for sequencing. *Arabidopsis thaliana* was chosen for molecular studies, principally for its short generation time with low DNA value (Bennett, 1972). The remarkably low amount of repeated sequences and the unique sequences being present in one or a few copies per nuclear genome and as such suitability for chromosome walking are the factors for choice of *A. thaliana* as the first angiosperm for entire genome sequencing through international collaboration. *Oryza sativa* has also been chosen for genome sequencing, partly because of the low DNA value and its economic importance. Lastly, it has been claimed that knowledge of C value of DNA has practical significance in view of recent suggestions that this may play an important role in determining which plant and crop species may survive effects of a nuclear winter (Grime, 1996), and respond to ozone depletion (Bennett, 1987), and even to global warming (Grime, 1996).

Lately, it has been shown that the quantitative change in genomic DNA, either increase or decrease, has been influenced by transposons. The transposon can expand the genome with its repetitive DNA to a very significant extent. As such, this C value paradox is now regarded as one in which transposons play a very significant role. Further study shows that environment can influence transposon activity helping to adapt to environmental changes. It is being claimed that maize used retrotransposons with the aid of RNA to double the genome size within a short evolutionary time. In fact, the number of transposons in maize has been found to be very high as compared to that of barley. Simultaneous to increase in genome size, there may be loss of transposons, leading to the decrease in size as well. Such cases often arise out of intrachromosomal recombinations, in which long terminal repeats (LTRs) undergo fusion with the excision of intervening DNA. The significance of the DNA increase has been much debated and discussed.

In general, small genomes with a fast cell cycle and short generation time have been considered to be of selective advantage. But in later results found in barley, retrotransposons show that DNA increases have a greater adaptive value. It has been noted that in barley plants, a type of retrotransposon that grows near the rim of a canyon is three times more redundant as compared to that at the bottom. It has been suggested that larger genome obtained through the ample presence of transposons may enable plants to grow in high and dry areas to influence the physiological system for retention of more water (vide Moffat, 2000).

CHAPTER 8

Sex Chromosome Differentiation

In the plant kingdom, sexuality is a late event in the scale of evolution. It is so far known that sexual reproduction, probably arising at the precambrian period, led to the origin of diverse forms of both plants and animals in nature. The diversity was dependent, to a great extent, on gene and chromosome mutation and sexual reproduction provided adequate scope for recombination.

With regard to bisexuality, the oldest landplants bearing the male and female reproductive organs are the late Devonian lycophytes (Cleal, 1995). But the oldest record of heterogamy and bisexuality is found in the green alga, the Silurian charophytes (Feist and Feist, 1997).

With the gradual evolution of sexuality, there has been differentiation of sex as well. The sex differences could be observed in the morphology, physiology, and other characteristics, all being controlled at the chromosomal level through their gene component. In the plant groups, there has been maximum, bisexuality in majority of species of flowering plants, and unisexuality only in certain families such as Euphorbiaceae, Cucurbitaceae, Araceae

and a few other taxa. In the unisexual plants, both male and female flowers may be present in a single plant as in monoecious species, whereas they may be located in different plants as in dioecious species. The presence of unisexual male and female flowers in the same plant as in species of Euphorbias is a matter of ontogeny and differential expression during development.

In case of flowering plants, the origin of dioecism from monoecism is rather clear or rather the step between the two is short, and in most cases is not well delineated. In many cases of dioecious species as in the genus *Mercurialis*, bisexual types also occur in nature. The species provides ideal material for the study of evolution of dioecism from hermaphroditism.

It has been suggested that at least two mutational steps have been involved in the origin of dioecism from a bisexual stage (Charlesworth, 1991). The two steps are: (1) a male sterility mutation converting bisexual to females and (2) the female sterility mutation converting bisexual into males (Figure 8.1). With the advent of unisexuality and dioecy, a primitive sex chromosome system was supposed to have

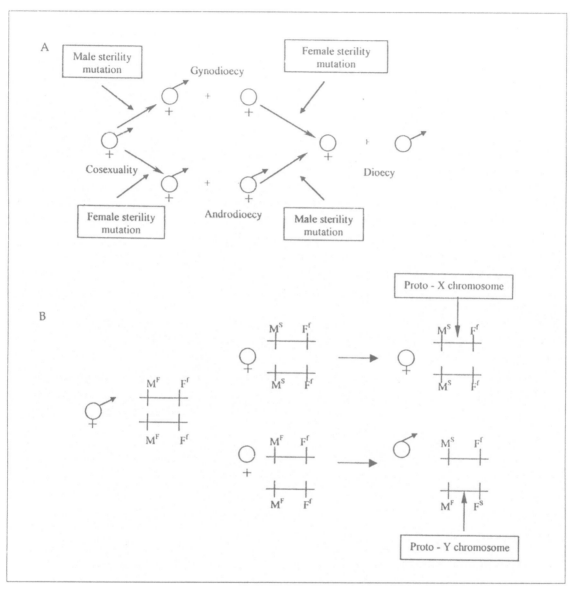

Figure 8.1   (A) The minimum number of mutational steps needed to produce dioecy. If a male sterility mutation arises first, a polymorphism for cosexuals and females (gyndioecy) is established. If a female sterility mutation arises first, a polymorphism for cosexuals and males (androdioecy) is established. Gyndioecy is not uncommon, but androdioecy is virtually unknown, as is predicted theoretically. (B) The evolution of proto-sex chromosomes, assuming an initial recessive male sterility mutation ($M^F \rightarrow M^S$) and a subsequent dominant female sterility mutation ($F^f \rightarrow F^s$) at a closely linked locus. Reproduced with permission from B. Charlesworth, *Science* 251: 1030–32. Copyright 1991 American Association for the Advancement of Science.

been established. The *proto X* would carry gene conferring female fertility and male sterility, and the *proto Y* female sterility and male fertility. This simple model suggests that initially evolution of sex chromosome involves restriction of recombination between genes controlling male and female function rather than throughout the entire chromosome. In the plant *Silene dioica*, the X and Y chromosomes are clearly heteromorphic. Region 1 of Y chromosome carries gene suppressing femaleness, region 2 promotes maleness, and region 3 promotes anther development (Fig. 8.2). The corresponding region of the X chromosome carries gene for femaleness and does not recombine with the Y chromosome. Because of their common ancestry, chromosomes X and Y show certain amount of homology. Gradual differentiation led to the development of X and Y chromosomes.

The genetic analysis of sex determination without visible structural differences may suggest a recent origin. Heteromorphic X and Y chromosomes are associated with the evolutionary advance and accumulation of repeat sequences. In *Melandrium album*, sex linked gene with degenerate Y linked homolog has been reported (Gutmann and Charlesworth, 1998). In *Carica papaya*, with the use of microsatellite probe, sex-specific differences have been worked out, where a putative Y chromosome has been reported having a sex-specific distribution (GATA) of microsatellite. Such cases are to be considered as intermediate step in the evolution of heteromorphicity from homomorphic state (Parasnis et al., 1999).

Only about 5% of the genera of higher plants are recorded to be wholly dioecious, while about 75% of the families have a few dioecious species. A survey of the Indian species shows that nine families comprising 140 genera and 939 species of the Indian flora are completely dioecious. Dioecism is presumed to have arisen independently in different plant taxa and the origin of cytologically distinguishable sex chromo-

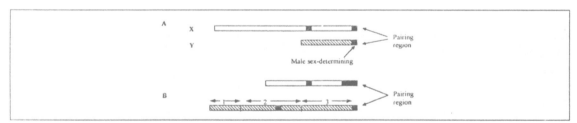

Figure 8.2 (A) Diagram of the general structure of the human X and Y chromosomes. The filled circles represent the centromeres. The unshaded region of the X contains active genes that are absent from the hatched region of the Y, with which it does not undergo recombinational exchange. The filled regions at the ends of the short arms of the X and Y undergo pairing and genetic exchange. The male-determining genes of the Y are located just to the left of the pairing region. (B) The structure of the X and Y chromosomes of the red campion *Silene dioica*. Region 1 of the Y carries a gene or genes suppressing femaleness, region 2 promotes maleness, and region 3 promotes anther development. The clear region of the X promotes femaleness and does not cross over with the Y. Reproduced with permission from B. Charlesworth, *Science* 251: 1030–32. Copyright 1991 American Association for the Advancement of Science.

somes might have been a gradual process. Such chromosomes may be absent in groups where dioecism is possibly a comparatively recent event. Dioecism at the familial level, such as Salicaceae and Menispermaceae, is rather rare. Their occurrence at the generic and subgeneric level such as *Humulus* and *Rumex* is also infrequent. Dioecism in flowering plants is confined, in general, to species or small groups of related species. The small size of sex chromosomes in some dioecious species has led to conflicting reports on the occurrence of sex chromosomes. In plants, all these evidences showed that the step between dioecism and bisexuality is rather short. As such, plant species provide scope for the study of evolution of dioecism from hermaphroditism or monoecism or vice versa.

Sex Determination in Bryophytes

Sex chromosomes in plants were first reported in the liverworts *Sphaerocarpus donnellii* Aust. (Allen, 1917), in which a large X in female and a much smaller Y in the male gametophyte were observed. The distinct morphological disparity of the heterochromatic chromosomes contributed by two sexes in the sporophyte was clearly brought out. Since then, chromosomes in mosses have been reported in a large number of species.

Bryophytes are characterized, in general, by the presence of two predominantly heterochromatic chromosomes forming the largest and smallest members (Smith, 1983) referred to as H and h chromosomes, respectively. It is usually the "H" that is differentiated into sex chromosomes, i.e. as X in the female and Y in the male gametophyte, though there are a few reports of

"h" differentiating into X and Y chromosomes as in *Pellia fabronians*, *Marchantia polymorpha*, *M. decipiens*, and *Conocephalus supradecompositus*.

If the autosomal complement is balanced, the sporophyte is normal, even when the X and Ys are present in abnormal ratios. This behavior may suggest that sex chromosomes are of little significance in the functioning of diploid generation. It is stated that the X chromosomes of *Sphaerocarpus donnellii* bear a gene or genes favoring femaleness, as plants with demaged X chromosomes may be male even in the absence of a Y chromosome. Moreover, the gene for femaleness is presumed to be near the centromere of the X chromosome, as plants of *S. donnellii* remain female even after the deletion of distal portions of the X chromosome.

In other liverworts, the so-called sex chromosomes may not show difference in morphology from the rest, or heterochromaticity, though the size may be different. Smith (1978), therefore, considered that the term "sex-associated chromosome" is preferable to sex chromosome.

In haploid gametophytes, there is usually only one sex-associated chromosome. However, in *Frullania*, there are two X chromosomes in the female and a single Y in the male. Usually, the sex-associated chromosome in hepatics is the largest in a few complements despite a few exceptions. The Y in *Sphaerocarpus donnellii* is an "m" chromosome. In *Pellia endiwifolia*, the sex-associated chromosomes are the smallest of the complement. Sex chromosomes as such may represent the largest pair of bivalents as in *Pellia neesiana* or may be smallest pair as in *P. endiwifolia*.

In mosses, heteromorphic type of bivalent has been recorded in several species and heteropycnosity in mitosis is distinct, as in species of *Bryum paradoxum* and *Hyophila involuta* (Verma and Kumar, 1980). Majority of polyploid species of mosses are monoecious, and it has been suggested that polyploids tending to change dioecism to monoecism, are associated with promotion of self-fertilization (Kumar, 1983).

Thus, in Bryophytes, the general trend of evolution is monoecism to dioecism, but a reversal of this process may occur through genome duplication. Regarding evolution of differentiated sex chromosome, a distinct trend has been noted. This trend indicates that evolution of sex chromosomes from completely undifferentiated X and Y to types with different amounts of heterochromatin and finally to morphological differentiation has been the trend in progress. As such, *Phagiochila primosa* represents the intermediate stage, as the two large chromosomes differ only in the amount of heterochromatin.

Sex Determination in Angiosperms

Sex chromosomes have been reported only in a few angiosperms, though there are several dioecious species.

Melandrium and *Coccinia*

Earlier studies on *Coccinia indica* Wight and Arn. (Syn. *C. cordifolia* L. and *Cephalandra indica* Naud.) with $2n = 24$ chromosomes indicated a pair of XY chromosomes in male, which was confirmed by later reports (Roy, 1974).

The somatic complements show 2A + XX and 2A + XY constitutions for females and males, respectively. The Y chromosome is conspicuously longer than the rest.

Aneuploids and euploids too have been raised (Chakravorty, 1948; Roy, 1974). The Y chromosome is fairly strong in determining maleness as the plants with the chromosomal constitutions of 3A + XXY (triploids), 3A + XXXY (triploid tetrasomic), and 4A + XXXY (tetraploids) are all males.

The fully fertile male organs formed in the subgynoecious plant, reported later, indicate that the basic sex trigger in *Coccinia* resembles that of *Melandrium* with the Y chromosome possessing a dominant female suppressor gene Su^F and at least two more linked genes M_1 and M_2 (Roy, 1974). The M_1 gene may be responsible for controlling the development of the male sex organ, such as the filament and the anther, while the M_2, comparable to the M_7 gene of *Melandrium*, controls the formation and fertility of the pollen. The X chromosome carries the alleles of Su^F, M_1, and M_2 in the differential segments.

Studies on *Coccinia* imply that the male-determining characteristic of the Y chromosome is comparable to that of *Melandrium* in the sense that the male-promoting effect of Y is stronger than the female-promoting effect of the X chromosome. In both *Coccinia* and *Melandrium*, plants with a single Y chromosome develop into males even in presence of three X chromosomes.

Chattopadhyay and Sharma (1991) observed that the difference between the male and the female plants in 4C nuclear DNA value, i.e. 10.35 and 8.25 pg in *Coccinia* is also due to the presence of long chromosome in the male plant. The heterochromatic nature of the Y chromosome is also indicated in its staining behavior. Difference in DNA value also shows high

DNA content in male due to large Y as compared to that in female.

The normal mechanism of sex determination in *Melandrium* is XX/XY (Warmke and Blakeslee, 1940; Westergaard, 1940), the staminate plants being heterogametic. In this genus, the influence of one Y chromosome is strong enough to suppress the female potency of three X chromosomes and four autosomal sets. However, the presence of one more X chromosome (XXXXY) results in bisexual and male flowers. They are mostly bisexual in Warmke's strains (1946) and mostly pure males or only slightly hermaphroditic in Westergaard's (1958) strains. Warmke noted that two Y chromosomes result in almost doubling of the male effect, balancing the effect of X chromosome and thereby majority of plants were male. The ratio of X/Y chromosome number was used as a parameter for measuring changes from complete male to hermaphrodic types. In Warmke's (1946) materials, all with a Y were males and all without the Y were females, implying no visible role of autosomes in sex determination.

The role of the autosomes, however, is evident (Westergaard, 1958) from the variation in sex expression in plants with the same sex chromosome constitution but with different autosome combinations. The percentage of hermaphrodites is more with the increased dosage of X and the percentage of male, with the increasing dosage of Y chromosomes.

The effects of sex chromosomes and autosomes in *Melandrium* have been quantified. A value of 1 for the male, 2 for the hermaphrodite, and 3 for the female has been assigned.

On the basis of X-ray induced breaks, the functional segments in Y chromosomes have been localized. Three different regions of the Y chromosome, each with a separate function in sex determination, have been identified. The absence of the distal part (the Y^1 type) leads to a normal bisexual plant, suggesting its role in suppression of the female sex organs. The loss of a part of the other arm (the Y^3) type develops a sterile male, implying that this segment must include gene or genes controlling the last stages in anther development. The absence of the Y chromosome (in XX plants) results in a female plant. Evidently, the middle region of the Y chromosome must include gene or genes that control the initiation of anther formation (Westergaard, 1958) (cf. Fig. 8.2). In this species (Dronamraju, 1965), a linkage is also claimed between a female suppressor segment and two blocks of male sex genes in Y chromosome. The X chromosomes and certain autosomes have female tendencies which are manifested specially in polyploids and aneuploids, with variable ratios of sex chromosomes and autosomes.

Rumex

This genus has been divided into four subgenera. Of these, *Acetosa* and *Acetosella* include all the dioecious species.

In *Acetosa*, sex chromosomes in higher plants were first discovered (Kihara and Ono, 1923) with the female having XX mechanism. The male plant is characterized by tripartite sex chromosome complex, consisting of one X chromosome and two Y chromosomes, which are unrelated but have affinities with opposite ends of the X chromosome.

The genus *Acetosa* is made up of several sections (Löve and Evenson, 1967) and the typical section *Acetosa* is subdivided into three subsections, *Acetosa, Insectivalves,* and *Americanae*. Females of *Acetosa pratensis* have 14 (12 + XX), whereas the males have 15 (12+XY$_1$Y$_2$) with three sex chromosomes.

The third subsection *"Americanae"* includes *A. hastatulus*. The main race of this species has only eight somatic chromosomes in the female and nine in the male plant. In male, there are three pairs of autosomes and XYY, similar to those of other *Acetosa* subsections forming the typical tripartite complex at meiosis. Another race is characterized by 10 chromosomes in both males and females with eight autosomes and XX:YY mechanism. Cytogenetics of the sex mechanism of this species (Smith, 1967) is presumed to be intermediate between that of *Acetosa* and *Acetosella*, since male factors may be carried on the Y chromosomes, and there is instability where the two races meet.

Section Paucifoliae of the genus *Acetosa* differs in morphology from that of the others. It has also two species, namely *A. paucifolia* and *A. gracilescens*. The former is diploid with 14 somatic chromosomes, i.e. 12 + XX/XY, whereas the latter is tetraploid with 28 chromosomes, in which one Y occurs in all males.

The two sections taken together clearly indicate that different sex chromosomal mechanisms have evolved for the determination of sex. The strong sex-determining factor of Y is pronounced. With increasing gene dosage or polyploidy, the reversal to autosomal stage or suppression of sex-determining property of sex chromosomes is rather significant. It is an index of sex chromosome–autosome interaction.

Cannabis sativa

The wild *Cannabis* is dioecious, but a variety of sex forms exists in the cultivated dioecious strains. Sex expression is influenced by environment, and the sex ratio may vary considerably from one strain to another.

Dioecious hemp has the normal XY mechanism with male heterogamety. Sengbusch suggested that all plants belonging to the monoecious or "uniform" varieties have the constitution XX, the variation in sex expression being due to heterozygosity of sex genes in the X chromosomes and in autosomes.

In *Cannabis sativa*, induced reversal of sex is also reported (Sarathi and Mohan Ram, 1974) in certain strains. Following treatment with gibberellic acid or silver nitrate, maleness could be induced in plants producing female flowers. With ethephon, the reverse process was achieved.

Chromosome analysis and estimation of nuclear DNA of both sexes of *Cannabis sativa* have been carried out (Chattopadhyay and Sharma, 1991) and the same chromosome number has been observed in both sexes. A single large Y chromosome was noticed in the male plant in the strain studied. In the female plant, the heteromorphicity is absent and as such it has two homomorphic chromosomes along with 18 autosomes. No clear heteromorphic bivalent could be noticed during meiosis and the chromosomes do not show strong heteropycnosity.

On the basis of the induced sex reversal in certain strains (Sarathi and Mohan Ram, 1974) and presence of heteromorphic chromosomes in others, it is possible that strains of this species may differ in the presence or absence of heteromorphicity in the male plants.

Humulus

Heteromorphic sex chromosomes with simple X–Y mechanism in *H. japonicus* and *H. lupulus* were identified in 1923. Later, Jacobsen (1957) recorded XY_1Y_2 mechanism in *H. japonicus*. In *H. lupulus*, the X is longer than Y, while in *H. japonicus* the three sex chromosomes are longer than the autosome. The Y chromosomes are partially heterochromatic.

Tiliacora

In *Tiliacora racemosa*, the two sexes show certain differences from each other. In somatic cells of the male, there is one large chromosome having no clear homolog in the karyotype. No positive or negative heteropycnosity could be recorded. The presence and absence of this long chromosome in male and female plant respectively suggest the possible existence of XY chromosomal mechanism in *Tiliacora racemosa*.

Sex chromosomal mechanism of sex determination in dioecious species of higher plants thus reveals conflicting results in a number of taxa. Major fraction of these species *does* not have detectable differences in chromosome complements associated with unisexuality. The sex chromosome may differ from the autosomes in specific finer details, not yet resolved.

Other genera

The unisexual species characterized by true homomorphicity is represented in species such as *Carica papaya*. The reversal of sex through decapitation is a common feature and the manipulation of sex through chemical treatments is well known in this species. In dioecious species of this category, evidently, both the male and female determiners are present and the expression is dependent on the repression and derepression of genes in the ontogeny of the organism. In this process of regulation, the autosomes too play an important role. However, recent findings on sex-specific differences on the basis of the microsatellite $(GATA)_4$ distribution need further analysis (Parasins et al., 1999).

In species of *Cannabis* as discussed earlier, both monoecious and dioecious forms are present. The occurrence of different sexual forms at the intrageneric and even at intraspecific level suggests a close interrelationship. A delicate balance between the sex-determining genes and the physiological setup plays a crucial role in the process.

The tendency towards the differential localization of sex-determining genes gradually led to the development of heterogamety, which in majority of the species has been associated with the accretion of heterochromatin in male or female, depending on the fact, whether the male as in *Coccinia indica* or female as in *Fragaria* is heterogametic. The progress of heteromorphicity has also gradually been associated with the delineation of specific functional segments, such as female suppression and anther development in linked loci of sex chromosomes.

Well-developed heteromorphicity associated in an uncontroversial way with unisexuality is distinctly recorded in species *of Coccinia, Cannabis, Melandrium, Acetosa, Acetosella, Humulus,* and *Tiliacora.*

General Trends

In the plant system, an analysis of the sex-determining property in X and Y chromosomes shows, in general, the longer size of Y as compared to X and a very dominant male determining tendency over the femaleness of X chromosomes or autosomes.

Multiplication of overall gene dosages as in polyploidy often affects the sex-determining tendencies of sex chromosomes and specially of Y chromosome. The degree of expression of maleness of Y differs to a great extent in aneuploids and euploids, as in *Acetosella*. It is a manifestation of sex chromosome–autosomal interaction in the determination of sex.

Polygamous condition is also of frequent occurrence in plants. The two sexes may be in different plants, in the same plant, or may be represented in hermaphrodite flowers. Such varied sexual types occurring at an intrageneric and intraspecific level indicate that the pro-

gression of homomorphicity to heteromorphicity is just a short step in evolution. Evidently, only a few genes are involved in the triggering of sexual forms, which may undergo reversion under certain conditions. As a result, such progression or reversion does not necessarily require a longes period of evolution. In species where dioecism is deeply entrenched, other adaptive changes in evolution might have altered the genetic architecture to a significant extent.

There has undoubtedly been a tendency in evolution towards the localization of sex-determining genes in specific segments and chromosomes. Evidences suggest that this evolution of monoecious to dioecious forms without necessarily implying heteromorphicity was centered around only a few genes, and the event might have occurred in the recent past.

CHAPTER 9

Identification of Chromosome Segments

Banding Patterns of Chromosomes and Sister Chromatid Exchange

In the identification of chromosome segments, the study of linear *banding patterns* in the chromosomes has emerged as a powerful tool. Chromosomes of an individual genotype show specific patterns of longitudinal differentiation, which, in absence of gross structural differentiation, serve as identifying criteria of a species. In certain cases, even functionally differentiated regions of chromosome, such as ribosomal or nucleolar regions, have specific banding patterns.

Genesis and types of bands

Initially, Caspersson and his colleagues (1971) noted that treatment with quinacrine dyes revealed strongly fluorescent chromosome segments when viewed through ultraviolet light. They further recorded differentially fluorescent segments, following staining with various fluorochromes under ultraviolet light. The bands were termed as *Q* bands. A concomitant discovery showed that such fluorescent segments corresponded to giemsa-stained bands, obtained following a short saline treatment. Such bands at intercalary regions were referred to as *G* bands. Since then, G-banding pattern has been obtained following treatments with a variety of chemicals, including saline, NaOH/HCl prior to saline sodium citrate (SSC), and trypsin.

The protocol for molecular hybridization has also been used in the banding technique, based on the principle that single strands of RNA or DNA are able to recognize and pair with their complementary base sequence. Through renaturation of DNA complex, originally denatured by various treatments, the eukaryote genome has been shown to contain repetitive sequences of DNA in addition to having unique or a few copy sequences of DNA (Pardue and Gall, 1970). The degree of reannealing and the time taken, or *reassociation kinetics*, is an index of the degree of repeats present in the chromosome. Repetitive DNA sequences are found to be present in high amounts in certain segments of chromosomes, specially the centromeres. When this protocol for molecular hybridization, that is denaturation and reannealing, is followed by giemsa staining, intense positive bands are revealed in similar segments of chromosomes. Such giemsa-stained bands are termed as C bands, which indicate the presence of highly repeated sequences. Such bands invariably represent heterochromatic segments. For nucleolar regions, a special banding technique has been developed, the as N banding, which primarily involves acid extraction. Differentiation of heterochromatin by N banding from C banded centromeric sites in plants is also on record.

Another banding technique, the R banding, gives bands opposite to that of Q and G and is termed "reverse banding," which has been extensively used in animal system to locate active gene loci. It was initially obtained through controlled heating for euchromatic regions. The O banding technique (Lavania and Sharma, 1979) uses trypsin and acid treatment for plant systems, revealing intercalary bands through orcein staining. In general, fluorescent banding, if enhanced with counterstain, is much more effective than normal cytochemical staining with absorption.

The plant system, G, C and Q banding have been applied by various authors to identify chromosome regions (Vosa, 1973; Marks and Schweizer, 1974; Kakeda et al., 1991).

The brief protocol for molecular hybridization, that is, denaturation at the cytological level, when followed by renaturation and staining with different dyes, particularly giemsa, gives intensely positive reaction at similar segments of chromosomes which otherwise show repetitive DNA. This banding, following denaturation–renaturation and giemsa staining, is termed C banding.

Saline treatment, followed by giemsa staining, gave a set of patterns similar in relative staining intensities to the Q bands. Such bands, known as G bands, did not require any denaturation–renaturation. Since then, G banding has been obtained following treatments in a variety of chemicals and even mere heating.

Orcein staining after incubation in XSSC results in O banding in plants, which is similar in general to the G band pattern. In addition, several other banding techniques have been reported from time to time, following modifications of the similar procedures as discussed above, as applied to specific segments (Sharma and Sharma, 1999).

Several DNA-binding specific antibiotics such as chromomycin and mithramycin are used as fluorescent dyes. It is possible to stain chromosomes with ethylene blue, followed by DAPI (4-6-diamidino-2-phenylindole), an AT-specific fluorochrome. Counter-staining with nonfluorescent compounds such as methyl green with chromomycin, and actinomycin D with DAPI, can be done. The C bands appear bright with the former and pale with the latter. Similarly, ethidium bromide removes and replaces quinacrine from pale-staining regions. The application of *restriction enzymes* on a chromosome has also resulted in differential staining of segments, with consequent manifestation of bands. The *restriction RE bands* observed following fluorescence may indicate base composition in certain areas of the chromosomes. Such bands may identify different classes of proteins as well as their interaction with each other and DNA, if it is followed by giemsa staining. The recognition sites of restriction enzymes are known, such as *Hae* II for GGCC, *Hind* III for AAGCTT. The bands may, therefore, represent G-enriched and A-enriched intercalary regions. Such RE bands, induced by restriction endonucleases, can be correlated with distribution frequencies of recognition sites of the corresponding restriction endonucleases in DNA, even in highly organized chromatin (Kamisugi et al., 1992). There are also evidences to show that organization of chromatin may be a factor in the activity of these enzymes (vide Sharma and Sharma, 1994).

Chemical basis of banding patterns

The chemical basis of giemsa banding is derived primarily from the fact the G bands are comparable to Q bands after treatment with fluorescent compounds and observed under UV light. It has been shown that with quinacrine mustard, a fluorescent amino-acridine nucleus becomes intercalated within the double helix of DNA. The basic N_2 atoms form ionic bands with DNA phosphate, and alkylating side groups form covalent bonds with guanine DNA. The similarity between giemsa and fluorescent quinacrine reaction is also borne out by the guanine specificity of fluorescent analog of actinomycin D (Hecht et al., 1974). The brightly fluorescent Q and G bands are AT rich, as evidenced by fluorescence with anti-adenosine antiboides, which yield a pattern similar to Q and G bands (Miller et al., 1974).

While quinacrine mustard (QM) stains AT-rich sequences and becomes distinct, with acridine orange, the picture is reverse especially with GC-rich telomeres. Alcoholic extracts of several alkaloids of Papaveraceae also reveal quinacrine-like fluorescence. As both AT- and GC-specific bands are available, the combined application permits study of sequence complexity in chromosomes.

Notwithstanding the above evidences of preferential renaturation of repetitive sequences in DNA being responsible for C banding, which is due to disruption and subsequent reformation of chromatin, Stockert and Lisanti (1972) claimed the manifestation of C bands to be equal, in both single- and double-stranded DNA.

Importance of the DNA–protein linkage

Although dye intercalation and secondary alkylation have been suggested to be responsible for G banding, the involvement of acid protein is also indicated by the induction of bands through the use of trypsin and compounds rich in the SH group of proteins. Similarly, chelating agents and other proteolytic enzymes alter the banding pattern (Lavania and Sharma, 1979). G and C bands can be related in a sequential manner by the progressive extraction of protein components associated with DNA (Chattopadhyaya and Sharma, 1988). Equally relevant is the report of Comings (1974) showing alkali and/or saline treatment to be a necessity for C banding; it removes 50% of DNA. Repetitive sequences of centromeric DNA must therefore be packaged in a structure much more resistant to extraction than the arms.

Mild treatment with trypsin produces G bands whereas prolonged treatment produces C bands instead of G. It has been claimed that differences in the proteins associated with DNA may explain the two types of banding. Since acid extraction does not produce bands, it has been suggested that nonhistone protein binding is the crucial factor. Under normal conditions, the entire chromosome stains with giemsa, whereas removal of proteins from certain sites, i.e. nonband regions, results in staining in areas (bands) where the protein is still left.

Importance of the DNA–protein linkage, as in G banding, is also noted in orcein banding, where the pattern is the same. Orcein is an amphoteric dye that stains both DNA and protein, the primary reaction involving tertiary amines of the chromosomal polypeptide (Sen, 1965). The orcein stain is very suitable for somatic chromosomes in plants because of its capacity for staining proteins as well. Treatment with mixture of sodium chloride and SSC involves removal of proteins from certain sites, where possibly because of the presence of unique sequences, binding to the dye may be comparatively weak. Dye is retained as shown by bands at sites where the binding is comparatively strong, due to the compact and homogenous nature of repetitive DNA sequences. Increase in the duration of treatment leads to stepwise disappearance of the band, except the C bands. It may be considered as an index of the relative difference in binding of chromosomal protein to highly repeated and to less repeated sequences of DNA. Apparently, binding of proteins is governed by the nature of the repeats. The distribution of proteins, as indicated by orcein banding, varies in the segments of chromosomes, depending on the type of DNA to which they are attached, or more precisely, to DNA of different functions.

Sister chromatid exchange

The development of banding pattern technology has also led to the improvement of techniques for demonstration of sister chromatid exchanges in somatic cells.

The sister chromatid exchange (SCE) technique is based on the principle of semiconservative replication of chromatids through thymidine uptake and autoradiography (Taylor et al., 1957). In the first cycle, only one chromatid shows incorporation. In the SCE technique, bromodeoxyuridine (BrdU) is incorporated through one or two cycles followed by fluorochrome and fluorochrome-cum-Giemsa staining (Latt, 1974, vide Sharma and Sharma, 1980, 1994) or by the indirect immunofluorescent method (Yanagisawa et al., 1993). Because of the problem of incorporating BrdU in plant DNA, such reports in plant system are relatively few (Vosa, 1977). To some extent, this limitation has been overcome by fluorodeoxyuridine (FdU), which inhibits thymidine synthetase and thus thymidilic acid. Simultaneously, additional uridine needs to be added to keep RNA synthesis unhampered. The SCE technique, with giemsa staining, can clearly differentiate substituted chromatids from nonsubstituted ones; similarly, brdU–giemsa staining can also differentiate late replicating DNA by pale color or dot formation. Certain authors, however, suggest that protein modification is res-ponsible for differential staining with giemsa. UV irradiation and trypsin treat-ment can also induce differential disinte-gration of BrdU-substituted chromatids. However, the technique is extremely useful in the study of spontaneous exchanges as well as the effect of environ-mental mutagens on the plant system and is regarded as a very sensitive cytological method for detecting potential mutagenic and carcinogenic agents. Moreover, the use of indirect immunofluorescent method permits the study of cell cycle and synchronization.

The importance of locating DNA sequen-ces in chromosomes through banding is due to the fact that the methods represent a merger of cellular and molecular methodology. It is specially relevant, as in the eukaryotic system, regulatory mecha-nisms at the chromosome level are highly complex. It is initiated at the molecular level in chromosomes, and through a series of cellular reactions culminates in the expression at the organismic level. The limitations of classical techniques do not normally permit the study of the entire process as a whole and keep the domains of molecules, cells, and organism as distinct. Modern methods including in vitro techniques as well are generally removing the demarcations at structural, ultrastructural, and molecular levels. The banding technique, which enables visuali-zation of molecular details at the cellular vis-a-vis the microscopic level, is indeed a synthetic procedure.

Identification of chromosome segments – molecular approach

In situ hybridization (ISH) techniques developed more than a quarter of a century ago, independently by Pardue and Gall (1970) and John et al. (1969), allowed identification of specific nucleic acid sequences directly on morphologically preserved chromosomes.

The term *"Chromosome painting"* widely implies painting of differential chromo-some segments with sequence-specific probes and is principally based on the technique of in situ molecular hybridiza-tion. The identification of a gene at the chromosomal level is necessary to study its function, its relationship with other genes, as well as for isolation and dissection. It is essential for both structural and functional genomics.

Basic principle, prerequisites, and different approaches

Localization and mapping of functionally differentiated segments of chromosomes and gene loci have been greatly facilitated through the application of molecular hybridization technique at the chromo-some level. The knowledge of the structural complexity of chromosomes at the molecular level, and the sequence com-plexity, has been gathered through an analysis of annealing of complementary sequences following hybridization of RNA and DNA strands. The method of in situ molecular hybridization principally uses probe sequences tagged with radioisotopes or a chemical reporter. The initial step is denaturation of the target chromosomal DNA mostly in metaphase, to facilitate access of the probe to the target. This is followed by hybridization with the probe while complementary sequences undergo pairing. The hybridized sites are localized through either autoradiography or immunofluorescence, depending on the type of probe used as well as counter-staining with specific stains.

The main disadvantages of the isotopic ISH introduced by Rayburn and Gill (1985) in plants are the limited spatial resolution, extensive time required for autoradiography, limited shelf life, and safety measures required in handling radioisotopes. However, modification of nucleic acid probes with a stable non-radioactive label has helped to overcome the major obstacles that hinder the general applicability of in situ hybridization. Initially, nonradioactive ISH techniques were applied for detection of repetitive sequences or amplified sequences in polytene chromosomes only.

Of the two types of molecular hybridization, isotopic and nonisotopic, chromosome painting is based on the nonisotopic one, the underlying principle of which is to make the specific loci of DNA antigenic (Gosden and Lawson, 1994).

Availability of multicolor fluorescent detection systems and reprobing of microscopic preparations have further facilitated simultaneous detection of several sequences and their linear order in situ through fluorescence in situ hybridization (FISH).

The basic strategy underlying the in situ hybridization technique involves several steps. DNA sequences are first labeled to produce the probe. Target chromosome spreads (treated to denature DNA to a single-stranded condition) are bathed with a solution of small single-stranded pieces of known probe DNA. Complementary sequences of the probe and target are allowed to reanneal. Related sequences anneal preferentially; therefore, probe DNA binds selectively at related

cytological sites. After washing and incubation in fluorescent-labeled reporter molecules, a discrete fluorescent signal is visible at the site(s) of probe hybridization, which can be detected on the chromosome using fluorescence microscopy.

Analysis of the different gene blocks in chromosomes and segments of complex genome has been carried out through the development and use of probes with differential fluorescence. Such *multicolor techniques* have been applied to localize repeated DNA and other sequences on chromosomes (Kenton et al., 1993; Mukai et al., 1993). The study of multicolor preparations included under chromosome painting may be *direct* or *indirect*.

In the direct method, the detectable molecule (reporter) is bound directly to the nucleic acid probe so that hybridization at sites can be microscopically visualized immediately. In the indirect method, the probe is labeled with a reporter molecule that is rendered detectable only after immunocytochemical localization.

Indirect labeling of probes involves the use of compounds such as biotin as the primary labels and later streptavidins, which are used as a conjugate for signal generation system. Otherwise, antibody of a hapten is incorporated into the probe, the recognition of which leads to detection. Similarly, photobiotin is detected either through luminescence or fluorescence or alkaline phosphatase conjugate – colorimetrically with enzyme conjugates. In general, indirect methods use probes tagged with biotin, digoxigenin, and dinitrophenol (DNP) as reporter molecules, which are detected by fluorochrome-conjugated avidin or antibodies. The common fluorochromes are FITC (fluorescin-isothiocyanate), rhodamine, and AMCA (amino-4-methyl coumarin-3-acetic acid).

The direct method, on the other hand, is characterized by probe labeling with fluorochrome-labeled antibodies. In direct labeling, the signal generation system is directly attached to the probe, which is detected posthybridization with the enzyme and enzyme conjugate. The examples include *horseradish peroxidase*, where the detection is through high chemoluminescence.

For detection of hybridization at the chromosome level, the nature of the probe is an important factor – larger the probe, easier is the detection. However, with the gradual development of amplification method at the in situ level, even single-copy sequences are amenable to detection (Levi and Mattei, 1995). Probes for chromosome painting can be obtained by either (1) *flow sorting* or (2) *microdissection of chromosomes*.

Repetitive sequences often hamper the identification of complex DNA sequences. Because of their nonspecificity, these sequences hybridize with their corresponding sequences present in the genome. To overcome this problem, *chromosomal in situ suppression (CISS)* or *competitive in situ hybridization (CISH)* is adopted. The method involves preannealing the probe with appropriate competitor DNA, rendering the repetitive sequences, especially the dispersed sequences, unavailable for hybridization. Normally, competitive DNA is the total fragmented unlabeled genomic DNA, enriched with repetitive sequences of desired length. The probe can then be amplified through PCR.

Different methods have so far been developed to delineate sequences specific to chromosome, chromosome region or, even genome.

Fluorescence in situ hybridization-FISH

The initial approach of FISH with the sequences, regions, or chromosomes as probe and later the in situ hybridization with *total genome (GISH)* as probe laid the foundation for further refinements.

One of the important modifications of the FISH technique is *genomic in situ hybridization* (GISH), where the total DNA from the genome of one parent of a polyploid species or a hybrid is labeled as a probe.

Plants with a large complex genome with similarity of repeats initially presented difficulties in delineating the desired repeated sequences with specificity as compared to animal system, where painting of short unique sequences is carried out along with *chromosomal in situ suppression of fast reassociating repeats*. DNA probes derived from chromosomes of specific regions of chromosomes as in *Vicia faba* (field bean) or *Picea abies* (spruce) led to the labeling of almost all chromosome regions (Fuchs et al., 1994). In later years, use of comparatively smaller genomes, chromosome-specific repeats as in B chromosomes, and refinements in methods yielding stronger hybridization signals have made chromosome painting applicable to plants as well. To achieve accurate localization of desired sequences, different modifications of the basic technique of FISH have been devised in different organisms and for different chromosomal sequences.

Two crucial modifications made so far regarding identification of DNA sequences are the *primer mediated extension* and *amplification in situ*.

Identification of single-copy and high tandem repeats

In situ hybridization coupled with PCR in chromosomes (Mukai, 1996), has been a new approach for physically localizing gene sequences in chromosomes. It is specially useful for low or single-copy genes. The principal requisites are that preparation of cells should be such that the morphology is preserved, diffusion of primers and enzymes into intracellular sequences are allowed, and amplification is permitted.

The other strategy for the detection of tandem repeated sequences, in addition to low-copy sequences, is to carry out the *oligonucleotide primed in situ synthesis* (PRINS) (Koch et al., 1989). In contrast to FISH, where labeled primers are necessary, unlabeled primer is used in PRINS. The unlabeled oligonucleotide primers are annealed to the denatured target DNA and extension is carried out in presence of labeled nucleotides and DNA polymerase. This method has been found to be best suited for high-copy tandem repeats.

Identification on extended DNA fiber

Refinements of the technique involve the hybridization of probes to extended DNA fibers. It involves, in principle, nuclear lysis, release of DNA fibers from lysed nuclei, spreading the DNA on the surface of the slide, and hybridization of probes following the standard schedule for fluorescence detection (Fransz et al., 1996). Physical mapping at high resolution in *Arabidopsis thaliana* through fluorescence hybridization has been achieved. This method, combined with multicolor FISH, can extend the lower limit of resolu-

tion up to 0.7 kb. Repetitive and single-copy DNA fragments of different sizes can be mapped with high resolution at the molecular level.

Localization of different sequences and gene mapping

The in situ technique, as it stands, permits localization of the distribution of tandem repeats, dispersed repeats, and complete genes in the chromosome (Plate 9.1). Cloned probes can be used for detecting of repetitive or amplified single-copy sequences. Chromosome-specific sequences can be used in YAC vector as well and can be identified later in the chromosomes, by the in situ technique, as done for chromosome 21 of the human genome.

The multicolor combinations of three primary colors can lead to seven different combinations and can delineate translocations in hybrids (Mukai, 1996) (Plates 9.2, 9.3). With the use of cloned repetitive DNAs, clusters of homogenous major repeats at certain loci in chromosomes of the complement have been localized (Sen et al., 1999). This technique has thus become a powerful tool for gene mapping (Plates 9.4, 9.5).

To identify *abnormalities, translocations, insertions,* and *breakage points,* genome-specific dispersed probes are proving to be useful. The probes generated from aberrant chromosome and their hybridization in metaphase can indicate branch-points, deletion, translocation, as well as confirm the normal sequences in the chromosome. This method is termed as *"reverse chromosome painting"*. Aberrant chromosomes, however, need to be sorted through flow sorting, labeling, PCR amplification utilizing oligonucleotide primer, and hybridization.

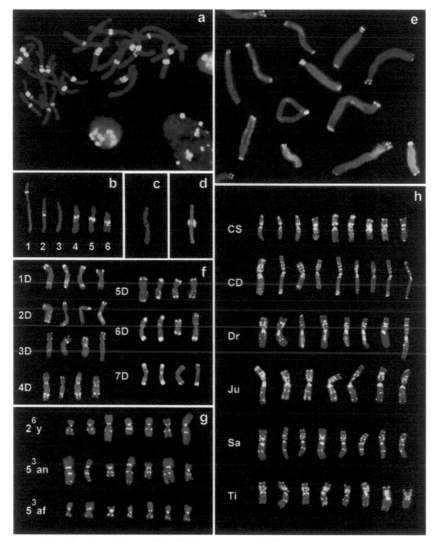

Plate   9.1   Localization of repetitive DNA sequences following in situ molecular hybridization

(a) Metaphase plate and interphase nuclei of field bean

(b) Localization of repeats on field bean chromosomes

(c) and (d) Visualization of repeats on field bean–acrocentric chromosomes

(e) Localization of telomeric repeats on field bean chromosomes

(f) Images of group D wheat chromosomes after localization of repeats

(g) Localization of GAA repeats in barley chromosome 5 isolated from two translocation lines and barley chromosomes 5^3 isolated from translocation line T3-5af

(h) Polymorphism in banding in wheat chromosome 3B

Reproduced by kind permission of Kluwer Academic Publishers from M. Kubalakova, J. Vrana, J. Cihalikova, M.A. Lysak & J. Dolezel, 2001: Localization of DNA sequences on plant chromosomes using PRINS and C-PRINS, *in*: Chromosome Painting, A.K. Sharma/Archana Sharma (Eds): a special issue of *Methods in Cell Science* **23**, 71-82, Kluwer Academic Publishers, Dordrecht. (The figure in question appeared on page 78 as Figure # 1).

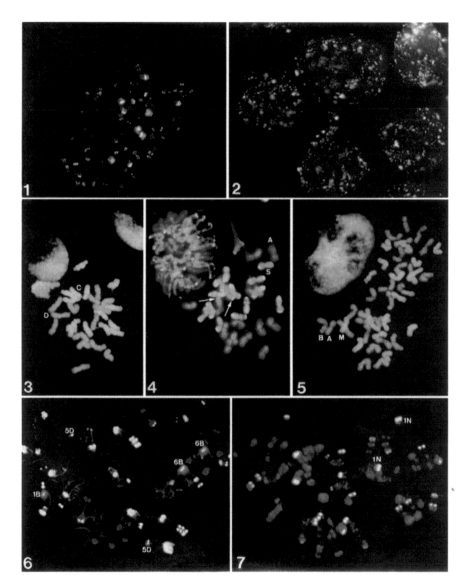

Plate 9.2 (1) – (2) Seven-color FISH on a metaphase cell (1) and interphase nuclei (2) of common wheat.

(3) Genome analysis of *Aegilops cylindrica* by multicolor FISH using total genomic DNA of its diploid progenitors as probes. The C- and D- genome chromosomes are detected by yellow and orange fluorescence, respectively.

(4) Simultaneous visualization of two genomes of synthetic amphidiploid (AASS). The A- and S- genome chromosomes are detected by orange and yellow fluorescence respectively.

(5) Simultaneous visualization of three genomes of a synthetic amphidiploid (AABBMM). The A-, B- and M- genome chromosomes are detected by orange, brown and yellow fluorescence respectively.

(6) Three color FISH of Chinese Spring wheat

(7) Genome allocation of the 18S, 26S rRNA genes of *Ae. ventricosa* using multicolor FISH. Y. Mukai (1996) "Methods of Genome Analysis in Plants"(Ed P. Jauhar). CRC Press, page 181–192. Reproduced by kind permission of CRC Press, USA

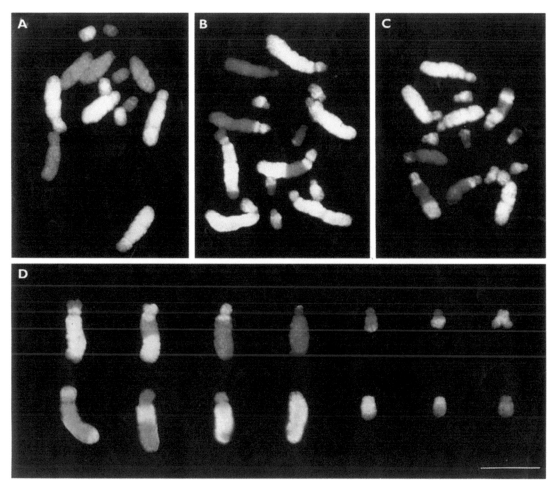

Plate 9.3 (A-D) Mitotic chromosomes of the *Gasteria lutzii* X *A. aristata* F₁ hybrid and back-crosses with *G. lutzii* following genomic in situ hybridization. The larger *Gasteria* genome (four long and three short chromosomes) fluoresces yellow; the smaller *Aloe* genome fluoresces orange. In each, the complete *Gasteria* genome fluoresces uniformly yellow while the recombinant chromosomes from the F₁ show yellow and orange segments, indicating cross-over points. C. Takahashi, I.J. Leitch, A. Ryan, M.D. Bennett and P.E. Brandham *Chromosoma* 105: 342-384 (1977). Reproduced by kind permission of Springer Verlag, GmbH & Co., Germany

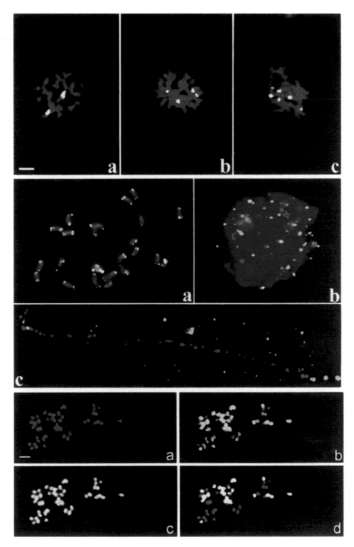

Plate 9.4 (1) 45S rDNA loci in the three representative *Oryza* species (a) *O. sativa* spp. *japonica*, (b) *O. sativa* spp. *javanica*, (c) *O. punctata*. *O. sativa* spp. *indica* has the same number of the loci as *javanica*.

(2) Multi color-FISH using rice A genome specific tandem repeated sequences (TrsA, red) and telomeric sequences (green).

(a) IR 36 has six pairs of TrsA sites at the distal ends of the long arm.
(b) Interphase mapping of TrsA and telomeric sequences.
(c) TrsA and telomere signals on the extended DNA fiber.

(3) Genomic in situ hybridization allows differential painting of C genome chromosomes among B and C genome chromosomes in *O. punctata* (BBCC).

(a) Counter stained chromosome with PI.
(Through the courtesy of Prof. K. Fukui, Hokuriku Agr. Expt. Station, Niigata, Japan. A.K. Sharma and A. Sharma (1999) "Plant Chromosomes - Analysis, Manipulation and Engineering". Harwood Academic. Reproduced by kind permission of Harwood Academic Publishers (Gordon & Breach), UK

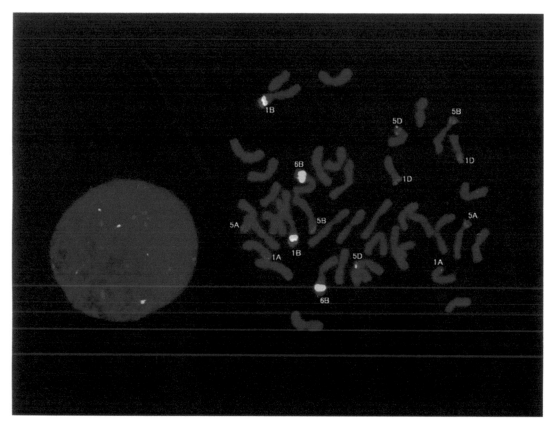

Plate 9.5 *In situ* hybridization - *T. aestivum* cv chinese spring 5s rRNA genes - 1A.1B, 3A, 5B, 5D (Red).
18S-26S rRNA genes - 1B, 6B, 1A, 5D (Green)
(Through the courtesy of Dr. Y Mukai, Osaka Kyoiku University, R-698- Asahigaoka Osaka, Japan)
A.K. Sharma and A. Sharma (1999) "Plant Chromosomes - Analysis, Manipulation and Engineering". Harwood Academic Publishers. Reproduced by kind permission of Harwood Academic Publishers (Gordon & Breach), UK

The in situ technique can be used to confirm *transgenesis*. The exact location of transgene sequences at the chromosomal level can be demarcated through chromosome painting. This is especially useful in genetically modified crops and animals, as successfully recorded in mice.

Assessment of genomic state in evolution

The type of material that can now be used readily as hybridization probe has expanded rapidly, ranging from small sequences of genomic DNA or cDNA through whole chromosome "paints" to the whole genomes used in comparative hybridization (Jiang and Gill, 1994).

Genetic maps are now available for almost all important crop species. One can now use these maps for marker-assisted selection in plant breeding program, for targeted gene cloning, and also to assist fundamental investigations of plant genome architecture.

The use of total genomic DNA as a probe for in situ hybridization shows the extensive differences in the genomes of related species.

Although metaphase analysis can help to distinguish chromosomes from the two parents in the hybrids where chromosomes are morphologically distinct, in most cases, owing to lack of karyomorphological markers, it becomes impossible to distinguish parental genomes. Schwarzacher et al. (1989) were the first to demonstrate the applicability of GISH in identification of parental genomes in an interspecific hybrid between *Hordeum chilense* and *Secale africanum*. Since then, this technique has been extensively used in several

systems to elucidate genome identification, their orientation, and spatial localization in interphase domains (Heslop-Harrison and Schwarzacher, 1996). Multicolor in situ hybridization has been used to distinguish three genomes in hexaploid wheat (Mukai et al., 1993) and three genome hybrids between wheat and grasses (Kosina and Heslop-Harrison, 1996).

In situ hybridization of the entire genome is also a very effective tool to elucidate ancestry of hybrids and polyploids. One of the most convincing cases is provided by the identification of ancestral genomes in the two common garden cultivars of *Crocus* cv. Stellaris ($2n = 2x = 10$) and cv. Golden yellow ($2n = 3x = 14$). The application of GISH, using differentially labeled genomic probes from the *C. flavus* ($2n = 8$) and *C. angustifolius* ($2n = 12$) clearly established that interspecific hybridization between them has contributed to the development of the said *Crocus* cultivars (Orgaard et al., 1996).

The ability of chromosomes to pair at meiosis is widely used to establish genomic relationships between species. Painting of chromosomes at meiosis permits determination of the genomic origin of paired and unpaired chromosome axis. In the hybrid between the hexaploid *Hordeum procerum* and tetraploid *Leymus racemosus*, pairing has been recorded between chromosome segments of *H. procerum* and *L. racemosus* at pachytene. However, at metaphase I, univalents and bivalents of single parental origin were found, while no *Hordeum–Leymus* bivalents were present (Heslop-Harrison, 1996). These data indicate the use of GISH in understanding intergenomic homologies as well as in elucidating the possible transfer of chromosome segments through inter-genomic recombination.

In general, in situ molecular hybridization aids in the construction of physical maps of the chromosomes. It can elucidate structural, functional, and evolutionary changes in chromosomes and delineate different genomes. It can indicate intra-chromosomal rearrangements and alien chromosome translocations as well as localize foreign genes in transgenesis.

Emerging areas

From basic and applied standpoints, two important applications of molecular hybridization at chromosome level need special mention.

The localization of isochore in the human system is a major achievement; it is holding good in plant system as well. Isochores (Saccone and Barnardi, 2001), which are GC-rich areas, form gene-rich segments, the localization of which facilitates genetic manipulation, especially in the identification of desired genes, their location, and finally their isolation. In the human system, general isochore families have been worked out, indicating areas of gene concentration. Such delineation clearly denotes the precise distribution areas of genes in the chromosomes. The understanding of this distribution can serve as a prerequisite for genetic engineering at specific sites, especially in transgenesis.

Finally, the similarity of gene order and colinearity between widely different organisms, termed *"synteny,"* especially in crop species, worked out a few years back, is paying high dividends in crop genetics, indicating the extent of genetic homo-geneity. Many genes are similar in most plants and it is clear that the ordering of genes is highly conserved across wide taxonomic groupings. For instance, the genome of wheat is 25-fold that of rice, but the linear order of markers is essentially the same in the homologous linkage groups in the two species (Moore et al., 1995) and in other genera as well. The entire family of grasses, with almost 10,000 species, maintains conserved colinearity of gene order. Syntenic arrange-ment has lately been found to be universal not only in major groups of plants but in the animal kingdom as well. The 19 building blocks of genes are considered to be basically responsible for the develop-ment of genomes of maize, rice, and grasses.

CHAPTER 10

External Agents in the Induction of Gene and Chromosome Alterations

The effects of outside agenes on chromosomes received progressive attention after the reports of X-ray induced mutations in *Drosophila* by Müller in 1927 and by Stadler (1928) in maize. These discoveries led to the observation of mutagenic properties of chemicals by Oehlkers in 1943 and Auerbach and her colleagues in 1946. A host of other workers led by Avery and Blakeslee (1959) used the colchicine to induce polyploidy in plants. Effects resembling those of radiation were referred to as *radiomimetic* effects and chemicals inducing them as radiomimetic chemicals (Sharma, 1984; Sharma and Sharma, 1960).

Effects of Chemicals

In the latter half of past century, increasing awareness of the importance of environmental factors has led to greater emphasis on prevention and to attempts to minimize human exposure to hazardous chemicals. There are thousands of natural substances to which humans are exposed and, in addition, manmade chemicals have become an integral part of modern society. Approximately 4.5 million chemicals are known to exist, the number growing by about 10% each year. Many of them have been identified for their mutagenic, clastogenic, carcinogenic, and teratogenic effects. Compounds producing chromosome abnormalities usually also induce gene mutations. Thus, the induction of chromosomal alterations may be taken to indicate genetic damage as well (see Fishbein, 1980, 1982; Sugimura et al., 1982).

The chemicals identified as carcinogens and mutagens represent a wide spectrum with a range of varying biological activities. Some of the common mutagens have been identified among food additives, hair colorants, and various industrial chemicals. Of the food additives, saccharin was reported to be occasionally carcinogenic, leading to its ban in the USA in 1977. Pesticides are widely used. Some of organochlorine and organo-phosphorus derivatives, such as DDT, have a long-term effect and enter the food chain. The effects of DDT and its metabolic products have led to a large amount of information being gathered on the genetical effect of these compounds and ban on some of them in the advanced countries, though these are still being used in the less advanced countries.

Some drugs have also been observed to induce chromosome damage in humans, including daunomycin. Amongst the stimulants, higher incidence of chromosomal aberrations has been reported in lymphocytes cultured from chronic alcoholics as compared to those from controls. The action of caffeine is more controversial, though a high intake has been observed to lead to a predisposition to loss of reproductive ability in women. In higher doses, nitrophenylinediamines, widely used in hair colorants, are in some cases found to be mutagenic or carcinogenic in mice. Mutagen and clastogen testing, therefore, has become an important part of toxicology (see Berg, 1979).

Genotoxic effect as reflected in carcinogenicity is the other facet of the problem. The relationship between mutagenic and carcinogenic effects is, however, not very simple, particularly since several compounds have been shown to behave as a carcinogen not necessarily through a direct action on DNA, but through their crucial role on growth and development of transformed cells.

Since the isolation of a pure carcinogenic chemical benzpyrene in 1932, more than 2500 chemicals with widely different structures, isolated from natural or environmental sources or synthesized, have been shown to be carcinogens and/or mutagens in different test systems. Although no single molecular feature has been pinpointed to be carcinogenic, yet intensive investigations have demonstrated that almost all chemical carcinogens, in their ultimate forms, are highly reactive electrophiles. Direct alkylating agents are electrophilic by nature and most other compounds undergo metabolic conversion to these reactive intermediates. These get covalently bound to tissue nucleophiles in the target organs and may initiate cancer or mutagenesis.

Electrophilic species generated from xenobiotics in vivo can be broadly classified into

a. nitrenium ions or nitrenes arising from functional moieties, nitro, nitroso, amino (primary, secondary and tertiary), acetamido, N-hydroxyl amino, etc.
b. epoxides from compounds, having double bond as polycyclic aromatic hydrocarbons, aflatoxins and amidopyrin, etc.
c. carbo-cations from N-nitroso compounds, cycasin, substituted hydrazines, etc.

A number of chemicals are known to induce chromosomal aberrations. They are also known as *clastogens*. In contrast to the X-rays, where such chromosomal changes can be detected almost immediately, the effects of these chemicals are delayed. They act on the G_1 and early S phases of the cell cycle and the appearance of the aberrations depends on DNA synthesis. Based on these two criteria, that is, the type of aberrations produced and the stage in the cell cycle affected, chemicals may be divided into four classes:

1. Chemicals producing gaps (achromatic lesions) and deletions after exposure in late S and G_2 phases. The biosynthesis of DNA and DNA precursors is inhibited. Relatively few compounds show this effect, including hydroxyurea, cytosine arabinoside, deoxyadenosine, and 5-fluoro-deoxyuridine.
2. Chemicals producing all types of chromatid aberrations after exposure in late S and G_2 phases. They are also few in number, such as 8-ethoxycaffeine and streptonigrin. Earlier workers classified their effects as nondelayed, radiomimetic, and independent of S-phase.
3. Chemicals producing all types of chromatid aberrations but only in cells treated in G_1 and early S. Most chemical clastogens belong to this group, including alkylating agents and nitroso compounds. The effect is delayed and is dependent on S phase for expression of aberrations.
4. Compounds which inhibit DNA repairs, such as caffeine.

Xenobiotics are metabolized in two stages, namely by biotransformation (phase 1 metabolism) and subsequently by conjugation (phase 2 metabolism). In the detoxication of carcinogens, or toxic chemicals, the oxidative reactions of phase 1 result in the formation of metabolites, which subsequently undergo conjugation of phase 2 by epoxide hydrolase, glutathione transferase, UDP-glucuronyl transferase, etc. to form conjugates, which are eliminated from the cell and finally the organism. In contrast, activation by oxidation results in the formation of proximate carcinogens or reactive intermediates.

The latter are usually poor substrates for the conjugating enzymes. Therefore, a nonenzymatic interaction of these reactive intermediates takes place with intracellular constituents, including proteins, RNA and DNA, leading to covalent binding and formation of neo-antigens, mutations, cancer, and cell death. These two alternative routes of oxidative biotransformation leading to detoxication or activation indicate two different modes of oxygenation and the probable existence of two families of enzymes, each responsible for the alternative pathways. The cytochromes P-450 result mainly in detoxication while the cytochromes P-448, flavoprotein monooxygenases, and non-enzymic free radical hydroxylations result in oxidative activations (see Tables 10.1 and 10.2).

Test Systems

Standard test systems have been devised to evaluate adverse long-term chronic health effects, such as cancer and heritable mutations, for chemicals of both everyday and rare use. Animal carcinogenicity tests are very expensive, time-consuming, and their sensitivity is limited. In testing for heritable germline mutations, the time required is about 2 years but a very large number of animals is used (up to 50,000) for a relatively low number of potential human mutagens.

These difficulties have led to numerous short-term techniques for detecting potential chemical carcinogens and mutagens, involving both prokaryotic and eukaryotic test systems. More than 200 such methods are available. Various agencies, such as the Environmental Protection Agency (EPA), International Commission for Protection

Table 10.1   Biological activation of xenobiotics (some examples)

Lethal synthesis	Fluoracetate	Fluorocitrate	Inhibition of aconitase and energy
Reductive activation	Mitomycin	Ring scission product	alkylation of DNA
	Azo compounds	Aromatic amines	
Activation of	Halocarbons	Halocarbon radicals	Activation of O_2 to O_2
oxygen	Quinones	Semiquinones	and formation of OH
	Nitro compounds	Nitroso radicals	
	Paraquat	Paraquat	
	(cation)	(radical)	
Oxidative	Hydrocarbons	Epoxides	Covalent binding to DNA and
activation		Quinones	proteins
C-oxygenation			
N-oxygenation	Amines	Hydroxylamines	Covalent binding,
			oxygen activation
S-oxygenation	Thio ethers	Sulfoxides	Phosphorylation
		Sulfones	of cholinesterase
Desulphuration	Thio-phosphates	Phosphates	
Activation			
by Conjugation	7-HMBA	HMBA sulfate	Covalent binding to DNA

Table 10.2   Major pathobiological injuries of chemical toxicity

Acute lethal	Inhibition of energy metabolism, ATP and Na+/K+ pump
Autooxidative	Lipid peroxidation, membrane disruption, DNA damage, and mutations
Immunological	Antigen formation, oxygen radical formation, inflammation

against Environmental Mutagens and Carcinogens (ICPEMC), and the World Health Organization (1985, *Guidelines for the Study of Genetic Effects in Human Populations*, WHO, Geneva) are advocating the use of such tests in assessing the carcinogenic and mutagenic potential of chemicals in humans. The principal endpoints for monitoring are:

1. mutational events, involving DNA, and
2. chromosomal aberrations, involving breaks and interchanges at different levels. These changes may occur in the somatic cells or in the germ cells. The methods include both in vivo and in vitro systems.

Tests with microorganisms are widely used in screening for mutagens due to the rapid growth rate of the cells, the large number of cells which can be tested and the technical simplicity and low cost of the tests. A widely used short-term test is the *Salmonella/Ames test* using several specially constructed strains of *Salmonella typhimurium* in culture for detection of frameshift or base pair substitution reverse mutations in auxotrophs. Rodent (or human) tissue homogenates (S9 mix) are employed to metabolize chemicals to their active mutagenic forms. They can detect most of the known organic chemicals with carcinogenic properties, but give negative reaction in testing for metal carcinogens.

Escherichia coli and *Bacillus subtilis* have also been used to identify mutations in auxotrophs and growth inhibition tests. *E. coli* K-12 is effective in inducing prophage and nucleic acid mutations. Of the lower eukaryotes, different strains of *Saccharomyces*, *Neurospora*, and *Aspergillus* are able to indicate mutations in culture, arising through gene conversion, forward and reverse mutations, mitotic recombination, and meiotic nondisjunction.

Established mammalian cell lines with stable karyotypes in culture have been employed to a limited extent in testing for mutagenesis at specific loci. Some common cell lines in use are Syrian hamster embryo, Chinese hamster ovary, and human diploid fibroblasts coupled to an external metabolic activation system. They are complementary to tests using the micro-organisms.

Several short-term tests have been developed to detect the ability of certain chemicals to interact with DNA by measuring how much DNA repair is induced. Since DNA repair decreases the DNA-damaging and mutagenic effects of chemicals, cells deficient in DNA repair can be more sensitive to lethal effects of chemical mutagens.

Testing schedules

Some of the more common methods utilising lower groups and plants are summarised below:

Bacteria

Salmonella typhimurium strains developed by Ames for use as testers for back mutation at the histidine locus constitute the most widely used and most useful systems for identifying genotoxic agents. The substance to be tested is added to several strains carrying different molecularly defined mutations in the histidine locus. Back mutations are identified. From the type of the strain exhibiting the presence of back mutations and from the conditions required to induce this response, it is possible to characterize the type of mechanism underlying the genotoxic action. It forms a sensitive qualitative probe for genotoxic activities of single chemicals or mixtures of different toxicants.

Bacterial strains are characterized by their specific and different mutations – base substitution or frameshift – in the histidine locus. In addition, all strains have genetic constitutions that make them more susceptible to penetration and attack by genotoxic agents.

Mutation *rfa* increases the susceptibility of the cell wall to penetration by substances added to the medium. Mutation uvrB blocks certain normal repair functions, which would ordinarily reduce the yield of back mutations. The plasmid *PKM101* enhances an error-prone repair mechanism. The amplification system, thus introduced again, increases the sensitivity of the parent strain to genotoxic damage. Strains have been engineered for specific purposes as well. Any response to the test treatment is dose-dependent. In the context of single identifiable substances, exposure is needed to a series of concentrations ranging from zero to slightly toxic levels.

In higher organisms, many substances nontoxic in the original form, acquire genotoxic properties during or after metabolic degradation. Classical examples of biodegradants with genotoxic properties are the degradation products of

DDT and malathion. These complexities may be simulated in vitro by adding a cell extract to the medium, containing the relevant enzyme complexes (*p*-450) together with the necessary energy sources, constituting the so-called S-9 mix. The cell extracts usually contain the microsomal fraction of mammalian (rat) liver from animals that have been pretreated with a substance, e.g. *Aroclor* 1254, which induces/stimulates the production of such degradation enzymes. This method has been further modified by TLC/Salmonella technique involving the separation of complex samples on thin layer chromatography and registering of mutagenic effects directly on the plates by the bacterium.

Yeast and fungi

Strains of yeast have been utilized in vitro as indicators of reverse mutations, gene conversion, mitotic recombination, aneuploidy, and other forms of chromosome damage. Growing cells of different species of *Saccharomyces* produce the enzymes necessary to activate many promutagens, which would not be able to induce mutations in vitro without S9 mix. Mutation studies have also been carried out with species of *Neurospora* and *Aspergillus*. These systems have a relatively limited scope.

Plants

Higher plants are used extensively for monitoring genotoxicity of chemicals due to the complexity of their genomes, their capacity in some cases to activate promutagens to mutagens and the ease of their use. The more commonly used ones are different species of *Vicia*, *Allium*, *Tradescantia*, *Hordeum*, and maize. For the localization of chromosomal alterations, karyotypes with marker chromosomes are required. Pretreatment of root tips with other chemicals prior to exposure alters the susceptibility in some cases.

Gene mutation studies have been carried out on *Arabidopsis*, *Glycine*, *Hordeum*, *Tradescantia*, and *Zea*.

Of these, *Zea mays* has been found to be a very good bioindicator because of the convenience of detecting mutation in the specific gene locus in identified chromosomes. The two tests that are widely employed involve chromosome 9.

(i) *Waxy locus pollen test.* The gene is located on the short arm of chromosome 9 and is responsible for the synthesis of carbohydrate amylose. The starch in individuals carrying the dominant allele *Wx* is a mixture of amylose and amylopectin, whereas amylose is absent in the double recessive individuals. Iodine test colors amylose blue-black; its absence gives a red or tan color. Since the pollen grain is haploid, containing either *Wx* or *wx*, the color can be used as monitor for this test.

(ii) *Yg-2 locus.* This gene is located on the short arm of chromosome 9 in maize and induces pale yellow green and slowly maturing leaves in homozygous recessive plants (Yg-2/yg-2). The dominant Yg-2 causes dark green coloration. In the heterozygous state, leaves are a normal green in color. In heterozygous individuals, mutation of the dominant allele leads to deficiency in the production of chlorophyll in the leaf sector. If a leaf primordial cell is involved, yellow green sectors develop in green leaf.

Micronucleus test

In early–early synthetic type of maize, the presence of micronuclei in root tips indicates chromosome aberrations. This test has been utilized in other plants, e.g. in root-tips of *Vicia faba* following exposure to radiation. Micronuclei are taken to indicate nondisjunction, resulting in whole chromosomes or acentric fragments being left out of the spindle.

Tradescantia staminal hair method
A change in color in high mutagen sensitive staminal hairs has been used to monitor air pollutants as well as physical and chemical mutagens. The mutational events change the color from blue to pink, which is observed in the dissected stamen under the microscope.

The *microspore* has been used to measure the extent of radiation as absorbed by the plant when exposed to radioactive cloud. In general, the microspores are excellent materials for scoring the effects of pollutants as shown by the formation of chromosomal aberrations during mitosis of pollen grains.

Vicia faba and species of *Hordeum* and *Arabidopsis* offer good systems for environmental monitoring. In the latter, due to the short life cycle, the effect of the pollutant can be measured in all phases of growth. Barley serves as an efficient environment monitor as well.

Allium cepa, Allium sativum test
This test, initially worked out by Levan (1949), is one of the most convenient methods of monitoring genotoxicity. The chromosomes of *Allium*, being quite large, allow a detailed analysis. The bulbs can be easily handled and roots with meristems grow out profusely at regular intervals at a rapid rate. Such regular and successive growth enables an analysis of both long- and short-term effects. The method involves the culturing of healthy bulbs in small beakers containing the solution to be tested with the growing roots immersed in the solution. The roots are cut at regular intervals to study the effects of different periods of treatment at different concentrations. The analysis of chromosome alterations is carried out following acidulated alcohol-fixation and orcein staining of the root-tips. Orcein staining has to be applied carefully as overheating may lead to chromosome breakage. To study the recovery from the toxic effects, bulbs with roots after treatment are transferred to nutrient medium for observation at successive periods.

Biochemical action

The analysis of mutagenesis has proceeded with the understanding of gene functioning. The mechanism for inducing mutation may involve either of the two phenomena: (1) directly induced base mispairing and (2) base misrepair. In the former, spontaneous or induced mispairing results in specific base pair substitutions. In the latter indirect method, mutations are one or more steps removed from the initial event. Alkylating agents give rise to many different reaction products in DNA, but only few of them (O^6-alkylalanine and O^4-alkylthymine) have been found to induce direct mispairing. Heat is an effective mutagen as well, particularly in large genomes. It converts cytosine to uracil and guanine to an analog of cytosine, leading to base mispair. When the elongation of a DNA chain is interrupted by the formation of lesions through the action of several products of alkylation, the postreplication repair process is often prone to errors due to the incorrect insertion of bases into gaps in the progeny-strand DNA opposite such a lesion.

Unlike mutagens inducing direct base mispairing, agents that trigger misrepair ultimately produce a wide variety of mutational types, such as transitions, transversions, frameshift mutations, and deletions.

DNA polymerases are possibly the most important enzymes of DNA synthesis, which influence the rate of spontaneous or induced mutations. These enzymes, being responsible for the fidelity of DNA replication, may either (1) select the incoming bases on the basis of instructions received from the parental bases or (2) select the bases first exclusively on the basis of their hydrogen bonding and then correct the misinsertions by "copyediting." Certain metals, known for their binding action with DNA polymerases, are thus found to be mutagenic. Manganese, for example, is highly mutagenic and stimulates the misincorporation of both ribose and deoxyribosenucleotides in culture systems.

Types of Effects

Effects on cell division

Phases of cell division affected by chemicals are: (1) the stage when the cells enter into division; (2) the initiation of spindle formation; and (3) cytokinesis.

Since DNA synthesis and oxidative phosphorylation are necessary in cell division, mitosis is usually inhibited by chemicals which affect these processes. For example, 5-fluorodeoxyuridine (FudR), deoxyadenosine (AdR), cytosine arabinoside, and aminopterin are capable of inhibiting mitosis due to their property of inhibiting DNA synthesis specifically. When the DNA synthesis is resumed through removal of the block, mitosis starts.

Respiratory inhibitors such as carbon monoxide and uncoupling agents such as dinitrophenol can also suppress cell division.

Agents inhibiting the first stage inhibit successively the division of the cell, the nucleus, and the chromosomes. The stage affected is interphase and occasionally early prophase.

Nature of effects

The effects on chromosomes may be divided into two categories: *reversible* and *irreversible*. In the former, certain characteristic features are observed: certain vital processes are switched off, but enough metabolic activities are retained, which permit the tissue to recover when the external stimulus is removed. Irreversible effects result either in instantaneous death or in toxicity leading ultimately to death of the tissue.

The observed reactions of the different chemicals can be divided into three main zones: lethal, narcotic, and subnarcotic (Levan 1949).

Lethal

Within the lethal zone, different kinds of effects are observed. The reactions are irreversible, leading to ultimate lethality in the tissue. The chemical may act as a fixative and preserve the vital structure, another chemical may give pycnotic clotting of the structure, and yet another a swelling up of the nucleus with dispersal of the nuclear matter all over the cell. Long treatment may result in the thinning of the chromosomes into residual threads. The extraction of nucleic acids by nonfixing chemicals has helped in studying the ultimate chromosomal threads. Heterochromatin is more resistant to the extraction of nucleic acid than euchromatin is, and centromeric heterochromatin more so than that in the other regions. With high pH and treatment with potassium cyanide, despiralization occurs and cells in division swell and die.

Narcotic

The main effects observed within the *narcotic* zone are C-mitosis or stathmokinesis and C-tumor formation. C-mitosis, so named because it was first observed with colchicine, takes place through the breakdown of the spindle after the chromatids have separated at the end of metaphase so that they lie within the same cell without subsequent cell plate formation. When the tissue is allowed to recover, the chromosome number is doubled, resulting in polyploidy. Prolonged treatment may lead to high degrees of polyploidy as observed with gammexane followed by multipolar stages on recovery. The C-mitotic activity is inversely proportional to its solubility in water in case of most chemicals. Colchicine is, however, an exception. It is highly soluble in water, but even very low concentrations (0.5%) are capable of causing spindle inhibition and arresting metaphase. As a result, a large number of metaphases can be obtained. C-tumor formation results in the formation of bead-like swellings in the root-tips. The cells, due to loss of polarity, undergo disorganized division. This effect may occur independently of C-mitosis, though it usually accompanies the latter. Gavauden divided C-mitotic chemicals into two groups: (1) those in which the threshold follows the physical property of the chemical, e.g. solubility, showing that the effect depends on a physical action, and (2) those in which a large margin is observed between reaction threshold and water solubility, indicating involvement of chemical reactions. An example is colchicine, in which one exchange of methoxy and aldehyde groups in the C rings forms iso-colchicine. The latter does not have C-mitotic activity (Vide Sharma and Sharma 1980).

Metaphase arrest has been observed with several other chemicals, such as acenaphthene, chloral hydrate, gammexane, etc. and can be used to analyze karyotypes.

Subnarcotic

Within the *subnarcotic* zone, several effects can be observed. However, the reaction is a reversible one, since the tissue, if removed from the chemical, recovers its normal activity after a certain period.

The different phenomena involved include (see Table 10.3)

a. *Pseudochiasmata:* Certain regions of mitotic anaphase chromosomes do not separate, showing bivalent-like configurations. The result is usually fragmentation and ultimate loss of the cell.

b. *Chromosome erosion:* Parts of the chromosomes are digested, resulting in a "starved" appearance.

c. *Fragmentation or chromosome breakage:* This can be caused by a wide range of chemicals. The mode of action is variable, as discussed earlier, though the chromosome component involved is finally DNA. The final upset of the nucleic acid metabolism results in hazards in protein

Table 10.3   Chromosomal aberrations

Term	Definition
Chromatid gap	An achromatic region in one chromatid, the size of which is equal to or smaller than the width of the chromatid
Chromatid break	An achromatic region in one chromatid, larger than the width of the chromatid. It may be either aligned or unaligned
Chromosome gap	Same as chromatid gap only in both chromatids
Chromosome break	Same as chromatid break only in both chromatids
Chromatid deletion	Deleted material from any part of chromatid
Chromatid fragment	A single chromatid without an evident centromere
Acentric fragment	Two aligned (parallel) chromatids without an evident centromere
Translocation	Obvious transfer of material between two or more chromosomes
Triradial	An abnormal arrangement of paired chromatids resulting in a triarmed configuration
Quadriradial	An abnormal arrangement of paired chromatids resulting in a four-armed configuration
Pulverized chromosome	A spread containing one fragmented or pulverized chromosome
Pulverized chromosomes	A spread containing two or more fragmented or pulverized chromosomes, but with some intact chromosomes still remaining
Pulverized cell	A cell in which all the chromosomes are totally fragmented
Complex rearrangement	An abnormal translocation figure that involves many chromosomes and is the result of several breaks and mispaired chromatids
Ring	A chromosome that is a result of telomeric deletions at both ends of the chromosome and the subsequent joining of the ends of the two chromosome arms
Minute (min)	A small chromosome that contains a centromere and does not belong to the karyotype
Polyploid or endo-reduplication	A cell in which the chromosome number is an even multiple of the haploid number (n) and is greater than $2n$
Hyperdiploid	A cell in which the chromosome number is greater than $2n + 1$ but is not an even multiple of n

reduplication, causing the chromosomes to break at different loci.

Fragmentation, followed by translocation, may lead to a new pattern of chromosome rearrangements, resulting in heritable phenotypic differences. The irradiation and chemotherapy of cancer is a principally associated with the excessive fragmentation of chromosomes through chemical agents (Koller 1945, 1946). Several chemicals break chromosomes at specific loci, showing the differential chemical nature of the segments; e.g. EOC (ethoxycaffeine) at nucleolar organizer and hydroxylamine at primary constriction regions.

d. *Translocations* are also observed as anaphase bridges.

These reactions usually do not have a threshold with the zone.

Some chemicals show all these three zones, others may show only one or two, depending on the concentration used. In maize, compounds related to mustard gas predominantly induce lethal changes, many of which are small deficiencies. The effect is often delayed, leading to mosaics or normal and mutant cells.

Certain organisms possess antimutagens, which counteract the mutagenic action of normal metabolic chemicals. Other chemicals, in combination, may give additive or synergistic effects.

Effects of Radiation

The effects of radiation on cells, and more particularly on chromosomes, form a very important aspect of study in the background of radioactive fallout and of X-ray therapy. After the first effect of irradiation observed by Müller in 1927 in inducing mutants of *Drosophila* through X-rays, this subject has assumed immense proportions.

Radiations may be ionizing or nonionizing. Ionizing radiations include X-rays, α, β, and γ-rays, fast neutrons, and the radiation emitted by radioactive substances. They act by the conversion of the molecules of the material through which they pass into electrically charged free radicals, such as $H^+ + OH^-$ from water, which are capable of ionizing further molecules. The intensity of an ionizing radiation is measured by the number of ionizations induced per unit matter, such as 1 *roentgen* being the intensity required to induce one million ionizations per 1 ml of matter. Compared to the total mass of a living organism, the dosage required for lethality is rather low, particularly in mammalian cells. Nonionizing radiations, such as UV rays and infrared rays, cause dissipation of the energy by molecular excitation within the tissue, the principal biological action being attributed to the differential absorption of energy by the cellular constituents, particularly nucleic acids.

The primary changes induced by X-rays include structural alterations of chromosomes following breakage and rejoining. The secondary effects arise from crossing between a normal and a primarily altered chromosome. Chromosomes at dividing stages are more susceptible to such changes, the meiotic chromosomes being more sensitive than mitotic ones. Breaks in the G_1 phase of the so-called "resting" stage lead to chromosome breaks while those in the S and G_2 phase result in chromatid breaks. Most chromosome breaks are restituted, only a minority forming new combinations. Restitution may be prevented partially if the material is centrifuged or exposed to supersonic waves after irradiation since it leads to a number of "wrong" recombinations and further structural aberrations. The same dosage given within a very short time shows a larger number of aberrations than when spread over a longer period since the breaks have enough time to heal.

The principal factors affecting radiosensitivity include both internal ones such as genotypes, chromosome type, size and number, stage of cell division, level of ploidy, and age of the tissue and external ones such as moisture, temperature, presence of oxygen and other atmospheric gases, dosage, storage, presence of other chemicals, and ionization density of the rays used. In yeast, more than 30 radiation-sensitive mutants have been located. They have different sensitivity to different DNA damaging agents, epistasis, effects on recombination, mutation rates, and sporulation. The *RAD 9* gene is seen to control cell cycle response to DNA damage.

The two theories regarding the mode of action of X-rays on chromosomes are the direct hit or target theory and the chemical theory.

The target theory, (Vide Lea, 1955) suggests that a direct hit by an electron may cause a gene to mutate, bringing about a chemical change with a different phenotypic expression or a chromosome region to break. It was originally supported by a linear relationship between the X-ray dosage and degree of gene mutation on the basis that a larger number of electrons would cause a larger number of hits. However, this relationship has not been observed in lower dosages and different species behave differently even with the same dosage.

The alternative *chemical theory*, given by Koller and his colleagues, holds that the X-rays possibly ionize the water in the cytoplasm, forming HO_2 and other precursors of H_2O_2, which are highly unstable oxidizing chemicals and act later on chromosomes by oxidation. It was validated by the demonstration that on transplanting a normal nucleus into an irradiated enucleated cytoplasm, breaks are induced in the chromosomes showing that the X-ray effect was mediated by the cytoplasm. Similarly unirradiated bacteria placed in an irradiated medium show an increased frequency of mutations. An additional evidence is the production of similar radiomimetic effects by both chemicals and X-rays.

The steps in the action of X-rays involve the breakage of the –SH group of protein, accompanied by the breakage of the DNA molecule mainly through the H^+ and OH^- ions and involving successively the removal of $-NH_2$ and –OH groups from the bases; oxidation specially of the pyrimidine bases so that even the ring is destroyed in some cases; and oxidation of the sugar, due to which the phosphate–sugar link is broken, resulting in a breakage on the nucleotide.

Some mutagens are found to be able to alkylate NH_2, OH, and phosphate groups. The breakage of DNA is due to the destruction of the H bonds between $-NH_2$ and –OH groups, which give DNA its rigidity by deamination and alkylation of NH_2. In the double helix model of DNA even a very small dosage may cause dealkylation and deamination at one point, duplicated in successive replications from the damaged nucleotide, thus explaining the large effects of even small doses.

On treating the tissue with certain substances such as cysteine or cysteamine before irradiation the lethal dose is increased, showing that these chemicals give protection against X-ray damage. Radiation damage is much reduced in the absence of O_2, possibly because it interferes with radical formation. Therefore, chemicals that act as oxygen acceptors or prevent formation of H_2O_2 or dissociation of H_2O may also act as protectors. Others such as carboxyl chloride and thiourea may protect the chromatid against direct hit by substituting –SH groups instead of proteins.

Beneficial effects of X-rays include gene mutations and structural alterations leading to better strains in agri- and horticulture. Investigations of factors governing radiosensitivity help in understanding the mechanism of the effect of ionizing radiations, in getting information useful for radiation protection and radiotherapy, and in inducing genetic variability in crop plants. Radiation can be used effectively in plant breeding if the frequency of induced mutations can be increased with decrease in chromosomal aberrations, physiological injury, and sterility and be directed or controlled.

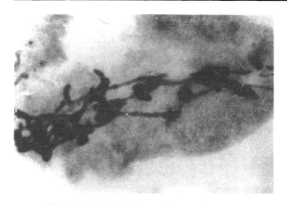

Photo 10.1 Stickiness in prophase

Photo 10.2 Chromosome bridge in anaphase

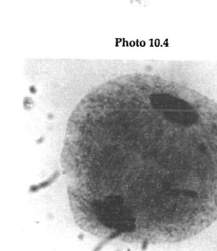

Photo 10.4

Photo 10.5

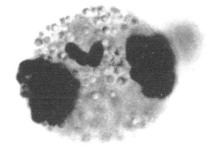

Photo 10.3

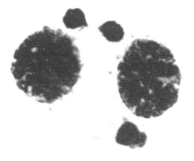

Photo 10.6

Photo 10.3 – 10.6 Lagging chromosomes and fragment in anaphase and telophase in meiotic cells.

Appendix A:
Some Simple Schedules for Study of Plant Chromosome*

STUDY OF MITOTIC CHROMOSOMES

(a) Staining in Feulgen
(b) Staining in aceto-orcein

Material: Root tips of *Aloe vera*

(a) *Staining in Feulgen*

  (i) *Pretreatment*: Keep freshly cut young root tips after washing in water in saturated aqueous solution of para-dichlorobenzene at 10–12°C for 3 h.
  (ii) *Washing*: Rinse in distilled water.
  (iii) *Fixation*: Transfer to acetic ethanol (1:2/1:3) and keep for 1–24 h.
  (iv) Bring down to water, passing through successively diluted grades of ethanol.
  (v) *Hydrolysis:* Hydrolyse the root tips in N.HCl at 58–60°C for 10 min.
  (vi) *Washing*: Rinse in water.
  (vii) *Staining*: Transfer the root tips to Feulgen solution, keep in cool place for 30 min to 1 h, till the tips are magenta colored.
  (viii) *Squashing*: Transfer each tip to a drop of 45% acetic acid on a slide, cut out the colored tip region, and discard the other tissue. Place a cover slip over the tip and squash applying uniform pressure with thumb on a piece of blotting paper.
  (ix) The preparation can be sealed with paraffin wax. Observe under the microscope.

---

* Some common plant species have been mentioned here. Certain variations of the schedule may be made for different materials after trials.

(b) *Staining in acetic-orcein*

Pretreatment and fixation are same as for (a)

(i) From the fixative, transfer the root tips to 45% acetic acid. Keep for 10–15 min.

(ii) *Staining*: Transfer the root tips to 2% aceto-orcein and N.HCl mixture in the proportion of 9:1, heat over a flame for nearly 5–10 sec, taking care that the liquid does not boil.

(iii) *Squashing*: Remove a root tip from the mixture and place it on a slide. Add a drop of 45% acetic acid on the root tip and take only the tip portion, remove the older part. Place a cover-slip on the stained tip part and squash by applying uniform pressure on the cover-slip with thumb through a piece of blotting paper.

(iv) *Observation*: Ring the cover-slip with paraffin wax and observe under the microscope.

From leaf tips

Leaf tip of Cestrum nocturnum

(i) *Pretreatment*: Dissect out very young leaf tips and immerse in saturated solution of para-dichlorobenzene for 3 h at 12–14°C.

(ii) *Fixation*: Fix the material in acetic-ethanol (1:2/1:3) mixture for at least 3 h, the period being extended, if necessary, up to 24 h to remove green color of the leaves.

(iii) Transfer to 45% acetic acid for 10–15 min.

(iv) *Staining*: Transfer the tips to a mixture of 2% aceto-orcein solution and N.HCl (9:1), heat over a flame for 5–10 sec, and leave the material in the stain at 30°C for 1 h.

(v) *Squashing*: Squash the tips on a drop of 1% aceto-orcein, cover with a cover slip, applying uniform pressure on a piece of blotting paper.

(vi) *Observation*: Seal with paraffin wax and observe under the microscope.

From pollen grain

Flower bud of Tradescantia paludosa

Dissect out an anther from a flower bud of suitable size, put a drop of 2% aceto-carmine solution on a slide, squeeze out the inner fluid, reject the empty anther lobes, smear with a clean scalpel, and cover with a cover-slip. Slightly warm over a flame and observe under the microscope. In flower of suitable size, mitotic division is observed in the pollen grains.

From endosperm

Endosperm of Cestrum nocturnum

(i) *Pretreatment:* Dissect out the very young developing seed under a dissecting microscope, place immediately in a saturated solution of aesculin, and keep at 10–12°C for 3 h.
(ii) *Fixation:* Fix in acetic-ethanol mixture (1:2) for 2 h. Transfer to 45% acetic acid.
(iii) *Staining:* Heat in a mixture of 2% aceto-orcein and N.HCl (9:1) over a flame for 5–10 sec, removing the tube at intervals so that fluid does not boil. Keep for 1 h.
(iv) *Squashing:* Transfer each seed to a clean slide in a drop of 1% aceto-orcein or 45% acetic acid, cut into two or three pieces with a scalpel, cover with a cover-slip and squash exerting uniform pressure. Seal with paraffin wax and observe.

STUDY OF MEIOTIC CHROMOSOMES

Meiotic chromosomes are studied usually from pollen mother cells and occasionally from embryo sac mother cells.

From pollen mother cell

Material: Flower bud of Rhoeo discolor

(i) Take flower buds serially from an inflorescence, starting from the smallest and working up to the largest, until the correct bud having divisional stages is found.
(ii) Dissect out a single anther from a bud with a needle. Place it on a clean slide. Add a drop of 2% aceto-carmine on the anther. Smear the anther with a clean scalpel. If needed, remove the debris. In case of large anthers such as of *Datura* or *Lilium*, cut the tip of the anther and squeeze out the inner fluid by pressing the other end. Keep the fluid on the slide and add a drop of aceto-carmine solution.
(iii) Heat slightly over a flame, cover with a cover-slip, and seal with paraffin wax.

Instead of fresh anthers, anthers fixed in acetic-ethanol mixture or Carnoy's fluid and later stored in 70% ethanol can also be observed following this method. Before smearing in aceto-carmine, keeping in 45% acetic acid is necessary.

From embryo sac mother cells

(i) Dissect out ovules from the ovary and fix in Carnoy's fluid for 1 day.
(ii) Keep in 95% ethanol for 1–2 days. Run through 90, 80, 70, 50, and 30% ethanol, keeping in each for 5–10 min. Rinse in water. Two different staining schedules can be followed.

Staining in Feulgen

    i) Hydrolyze for 8–10 min at 58–60°C in N.HCl.

   ii) Rinse in water and stain in Feulgen solution.

  iii) Transfer to a drop of 45% acetic acid on a clean slide and squash under cover-slip applying uniform pressure and observe.

Staining in aceto-orcein

    i) Transfer the ovules from water to 2% aceto-orcein and N.HCl mixture. Heat gently over the flames for 5–10 sec without boiling the fluid.

   ii) Keep in the stain for 20 min.

  iii) Transfer to a drop of 1% aceto-orcein on a clean slide and place a cover-slip and cover with a blotting paper. Squash under the cover-slip, exerting uniform pressure on blotting paper and observe.

SCHEDULE FOR BANDING PATTERN

Material: Root tips of *Allium cepa*

With orcein staining

The root tips after pretreatment and fixation:

    i) Hydrolyze for 20–25 sec in N.HCl at 60°C.

   ii) Squash in 45% acetic acid under a cover-slip.

  iii) Invert slide with cover-slip in absolute ethanol and keep till cover-slip is detached.

  iv) Treat both slide and cover-slip with material in a saturated solution of barium hydroxide for 5 min.

   v) Wash thoroughly in distilled water.

  vi) Incubate in 2X SSC at 60°C for 25 min to 1 h.

 vii) Wash in distilled water.

viii) Dry slide and cover-slip by holding in front of current of warm air.

  xi) Stain with 1% aceto-orcein and mount with a fresh cover-slip. Seal and observe.

Staining with Giemsa

    i) Pretreat the active growing root tips in 0.05% aqueous colchicine or saturated aqueous solution of para-dichlorobenzene for 3–4 h.

   ii) Fix in acetic acid-ethanol (1:3) for overnight.

  iii) Hydrolyze the roots in N.HCl at 58–60°C for 4–5 min.

  iv) Squash on a slide as usual.

   v) Detach the cover-slip from the slide.

  vi) Air dry the slides.

vii) Treat the slides with squashed material in saturated solution of barium hydroxide for 5 min.

viii) Wash in distilled water.

ix) Immerse in 2X SSC at 60°C for 45–60 min.

x) Rinse in distilled water.

xi) Stain in 5% Giemsa (E. Merck) at pH 6.8 in phosphate buffer for 20–30 min.

xii) Mount in Euparal.

Methods for Making Temporary Slides Permanent

i) Remove the paraffin seal.

ii) Invert the slide in ethanol and acetic acid mixture (1:1). The cover slip will detach from the slide.

iii) Transfer the slide and the cover-slip with attached material to ethyl alcohol and normal butyl alcohol (1:1) and keep for 5 min.

iv) Transfer the slide and the cover-slip to *n*-butyl alcohol for 10 min.

v) Transfer the slide and the cover-slip to another set of *n*-butyl alcohol for 30 min. If needed, the slides can be kept overnight.

vi) Mount in Euparal.

PREPARATION OF STAINS

1. Feulgen
2. Aceto-orcein
3. Aceto-carmine
4. Giemsa

Feulgen stain (Leucobasic fuchsin solution)

i) Take 0.5 g of basic fuchsin and dissolve in 100 ml of boiling distilled water in a conical flask.

ii) Cool down the solution at 58°C and filter into a dark colored bottle.

iii) Allow the temperature to come down to 26°C.

iv) Add 10 ml of N.HCl and 0.5–1 g of potassium metabisulphite. Cork the bottle tightly and keep in a cool place at 16–20°C for 24 h.
If necessary, seal the bottle with paraffin wax to make airtight.
Within this period, the dark magenta color solution becomes straw colored due to bleaching caused by the liberation of sulfur dioxide by action of HCl on metabisulphite.

v) If the solution does not turn straw colored after 24 h, add a pinch of charcoal powder, shake, make the bottle airtight, and keep in cold for 24 h. Filter and keep in a cool, dark place.

Preparation of aceto-orcein stain

Take 2 g of orcein and dissolve gradually in 100 ml of boiling 45% acetic acid. Heat the solution for 10 min carefully keeping it at a simmering point. Allow the solution to come down to room temperature and then filter.

Preparation of aceto-carmine stain

Take 1 g of carmine and dissolve it gradually in 100 ml of boiling 45% acetic acid. Heat the solution for 15 min carefully keeping it at a simmering point. Cool down the solution to room temperature and then filter.

Preparation of giemsa stock solution

 i) Dissolve 1 g of Giemsa (E. Merck) in 54 ml of glycerol.
 ii) Mix after cooling with 84 ml methanol and filter.

Commonly used pretreatment chemicals

Para-dichlorobenzene, 8-hydroxyquinoline, coumarin, colchicine, and aesculine for different periods.

Bibliography

Allen CE (1917) *Science* 46:466.

Avery AG, Satina S, Rietsema J, Blakeslee AF (1959) *"Blakeslee"* the genus *Datura*. Ronald Press, New York.

Bennett MD (1972) *Proc Royal Soc Lond B* 181:109

Bennett MD (1982) In Dover GA, Flavell A (Eds.), *Genome Evolution*, p. 239. Academic Press, London.

Bennett MD (1987) *New Phytologist* 106 (suppl): 177.

Bennett MD, Leitch IJ (1995) *Ann Bot* 76:113.

Bennett MD, Leitch IJ (1998) *Ann Bot* 82 (suppl A) 1.

Bennett MD, Smith JB (1991) *Phil Trans Royal Soc London* B 334:309.

Bennett ST, Leitch IJ, Bennett MD (1995) *Chromosome Res* 3:101.

Bennett R (1996) *Curr Opinion Genet Dev* 6:221.

Berg J (1979) (Ed.) *Genetic Damage in Man Caused by Environmental Agents*. Academic Press, New York.

Berrie GK (1974) *Bull Bot Soc Fr* 121:129.

Bhaduri PN, Bose PC (1947) *J Genet* 48:237.

Bhattacharya D, Sharma AK (1952) *Genetica* 26:410.

Blackburn EH (1990) *Science* 249:489.

Blackburn EH (1991) *Nature* 350:569.

Bloom W, Lewis H (1972) In CD Darlington, KR Lewis (Eds.), *Chromosome Today* 3, 268 pp., Longman Green, London.

Bodner AG, Guellette M, Forlkis M, Holt SE, Chiu CP, Morin GB, Harley CB, Shay HW, Lichsteiner S, Wright WE (1998) *Science* 279:349.

Brandham P (1983) *Kew Chromosome Conference*, Vol. 2, 251. George Allen and Unwin, London.

Brown SDM (1984) In Chopra VL, Joshi BC, Sharma RP, Bansal HC (Eds.), *Genetics: New Frontiers*, 221pp. Oxford & IBH, New Delhi.

Burnham CR (1962) *Discussion on Cytogenetics*. Burgess, Minneapolis, USA.

Camara A (1951) *Broteria* 20:1.

Camara A (1957) *Broteria* 20:5.

Carpenter BG, Baldwin JP, Bradbury EM, lbel K (1976) *Nucleic Acids Res* 3:1739.

Caspersson T, Lommakka G, Zech L (1971) *Hereditas* 67:89.

Cavalier-Smith T (1976) *Nature* 262:255.

Chakravorti AK (1948) *Proc Ind Acad Sci* B 27:74.

Chakravorti AK (1951) *Ind J Genet Plant Breed* 11:34.

Chal C (1995) In MJ Benton (Ed.) *Fossil Record* 2, pp. 779–794. Chapman and Hall, London.

Chandlee JM (1990) *Physiol Plant* 79:105.

Charlesworth B (1991) *Science* 251:1030.

Chattopadhyay D, Sharma AK (1988) *Staind Technol* 63:283.

Chattopadhyay D, Sharma AK (1991) *Feddes Rep* 102:29.

Chen PD, Fujimoto H, Gill BS (1994) *Theor Appl Genet* 5:135.

Cole KM (1990) In KM Cole, RG Sheatte (Eds.), *Biology of the Red Algae*, Cambridge University Press, USA.

Comings DE (1974) In H Busch (Ed.), *The Cell Nucleus* 1. Academic Press, New York.

Connell M, Nurse P (1994) *Current Opin Cell Biol* 6:867.

Cook PR (1973) *Nature* 245:23.

Creighton MR, McClintock B (1931) *Proc Natl Acad Sci* 17:485.

Crick F (1971) *Science* 204:264.

Darlington CD (1937) *Recent Advances in Cytology*. Blakiston, Philadelphia.

Darlington CD (1939) *J Genet* 37:341.

Darlington CD (1965) *Recent Advances in Cytology*, 2nd ed. Churchill, London.

Davidson EH (1976) In, *Gene Activity in Early Development*, Academic Press, New York.

Davidson EH, Britten RJ (1979) *Science* 204:1052.

Dennis ES, Peacock WJ (1984) In Chopra VL, Joshi BC, Sharma RP, Bansal HC (Eds.), *Genetics: New Frontiers*, p. 247. Oxford & IBH, New Delhi.

Doring HP, Tillman E, Starlinger P (1984) *Nature* 307:127.

Dronamraju KR (1965) *Adv Genet* 13:227

Egel R (1981) *Nature* 290:191

Essad S, Vallade J, Comu A (1975) *Caryologia* 28:207.

Evans LS, Van't Hof (1975) *Amer J Bot* 62:1060.

Fanti L, Dorer DR, Bertoco M, Henikoff S, Pimpinelli S (1998) *Chromosoma* 107:286.

Federoff NV (1984) *Sci Amer* 250:64.

Feist, Feist R (1997) *Nature* 385:4015.

Feldman M (1988) In TF Miller, RM Koebner (Eds.), *Proceedings of the International Wheat Genetics Symposium* 23, Cambridge.

Finchman JRS, Sastry GRF (1974) *Annu Rev Genet* 8:15.

Fishbein L (1980) *J Toxicol Env Hlth* 6:1133.

Fishbein L (1982) In T Sugimura, S Kondo, Takebe H (Eds.) *Environmental Mutagens and Carcinogens*, p. 307. Alan R Liss, New York.

Fisher D, Nurse P (1996) *EMBO J* 15:850.

Flavell A (1980) *Rev Physiol* 31:569.

Forsburg SL, Nurse P (1991) *Annu Rev Cell Biol* 1:227.

Fransz PF, Stam M, Montign BM, Hoopen RI, Wiegant J, Kooter JM, Oud'O Nanninga (1996) *Plant J* 9:767.

Fuchs J, Houben A, Brandes A, Schubert I (1996) *Chromosoma* 104:315.

Fuchs J, Pich U, Meiser A, Schubert I (1994) *Chromosome Res* 2:25.

Fuchs PU, Schubert I (1996) *Chromosome Res* 4:207.

Garber ED (1972) *Cytogenetics.* McGraw Hill, New York.

Geitler J (1953) *Endomitose und endomitolische polyploidisterung.* Springer-Verlag, Berlin.

Georgiev GP, Nedospasov SA, Bakayev. VV (1978) In H Busch (Ed.), *Cell Nucleus* 6, p. 3. Academic Press, New York.

Gerstel DU, Burns JA (1976) *Genetics* 46:151.

Gesteland RF, Atkins JF (1993) (Eds.) *The RNA World*, p. 632.

Gill BS, Friebe B (1998) *Current Opinion in Plant Biology* 1:109.

Gill BS, Friebe B, Endo TR (1991) *Genome* 34:230.

Gillies CB (1981) *Chromosoma* 83:575.

Gillies CB (1983) In PE Brandham, MD Bennett (Eds.), *Kew Chromosome Conference II,* 115 pp.. George Allen and Unwin, London.

Gosden JK, Lawson D (1994) *Human Mol Genet* 3:931.

Greilhuber J (1998) *Ann Bot 82* (Suppl A):27.

Greilhuber J, Ebert I (1994) *Genome* 37:646.

Grime (1996) *Aspect of Applied Biology* 45:3.

Guerra M, Kenton A, Bennett MD (1996) *Ann Bot* 78:157.

Gupta PK (2001) *Nucleus* 43:94.

Gupta PK, Priyadarshan PM (1980) *Adv Genet* 21:255.

Guttman DS, Charlesworth D (1998) *Nature* 393:263.

Hadlaczky G, Went M, Ringertz NR (1986) *Exp Cell Res* 167:1.

Harfman N, Bell AJ, Mcllanchlan (1979) *Biochem Biophys Acta* 564:372.

Hennig W (1999) *Chromosoma* 108:1.

Herbert A, Rich A (1999) *Nature Genetics* 21:265.

Heslop-Harrison JS (1996) *Unifying Plant Genome Comparison Colinearity and Conservation.* Company of Biologists, Cambridge.

Heslop-Harrison JS (2000) *The Plant Cell* 12:617.

Heslop-Harrison JS, Schwarzacher T (1996) In PP Jauhar (Ed.), *Methods in Genome Analysis in Plants.* 163 pp. CRC Press, Boca Raton, Florida.

Heumann JM (1976) *Nucleic Acids Res* 3:3167.

Hozier JC (1979) In JH Taylor (Ed.), *Molecular Genetics,* 3rd Ed. 315 pp. Academic Press, New York.

Huben A, Brandes A, Pich U, Mantenffel R, Schubert I (1996) *Theor Appl Genet* 93:477.

Huskins CL (1948) *Nature* 161:80.

Huskins CL, Steinitz L (1948) *J Hered* 39:66.

Ingle J, Timmis JN (1975) In P Markham (Ed.), *Modification of the Information of Plant Cells,* 37 pp. Elsevier, North Holland.

Innocenti AM (1975) *Caryologia* 28:225.

Iyer RD (1968) *Curr Sci* 37:181.

Izant JG, Weintraub H (1984) *Cell* 36:1000.

Jacobsen P (1957) *Hereditas* 43:357.

Jeffreys AJ, Royle NJ, Wilson V, Wong Z (1988) *Nature* 332:278.

Jiang T, Friebe B, Gill BS (1994) *Euphytica* 73:1994.

John HA, Birnstiel M, Jones KW (1969) *Nature* 223:282.

Kakeda K, Fukui K, Yamagata H (1991) *Theor Appl Genet* 81:144.

Kamisugi Y, Ikeda Y, Ohno M, Minezawa M, Fukui K (1992) *Genome* 35:793.

Kamptla SP, Chandrasekharan MB, Iyer LM, Li G, Hall TC (1998) *Trends in Plant Science, Reviews* 3:97.

Kenton A, Parokonny AS, Gleba NY, Bennett MD (1993) *Mol Gen Genet* 240:159.

Khesin RB (1985) Nauka, Moscow (In Russian).

Khush GS, Rick CM (1964) *Science* 145:1432.

Kihara H, Ono T (1923) *Bot Mag (Tokyo)* 37:147.

Kipling D, Fragsher GA (1999) *Nature* 198:191.

Koch JE, Kolvraa S, Petersen KV, Gregersen M, Bolund L (1989) *Chromosoma* 98:259.

Koller PC (1995) Nature 155:778.

Koller PC (1948) Nature 162:514.

Koltunow AM (1993) *The Plant Cell* 5:1425.

Kriegstien HJ, Hogness DS (1974) *Proc Natl Acad Sci USA* 71:135.

Kumar SS (1983) In *Genetical Research in India*. ICAR, New Delhi.

Laemmli UK et al. (1978) *Cold Spring Harbor Symp Quant Biol* 42:109.

Laskey RA, Earnshaw WC (1980) *Nature* 286:763.

Latt SA, Brodie S, Munroe SH (1974) *Chromosoma* 49:17.

Lavania UC (1999) *Curr Sci* 77:216.

Lavania UC, Sharma AK (1979) *Stain Techn* 54:261.

Lavania UC, Sharma AK (1984) *Experientia* 40:94.

Lee HW, Blasco MA, Gottlieb GJ, Homer JW II, Greider EW, De Pinho RA (1998) *Nature* 369:569.

Levan A (1949) Hereditas (Suppl vol) 326 Proc.

Lea DE (1955) *Action of radialation of cells*. Cambridge. The University Press.

Levi N, Mattei M (1995) In Eds. BP Hames and SJ Higgins *Gene Probes 2: A Practical Approach*, 212 pp. IRL Press, Oxford.

Lewin B (1980) In *Gene Expression–Eukaryotic genomes*. Wiley-Interscience, New York.

Lieckfeldt E, Meyer W, Bramer T (1993) *J Basic Microbiol* 33:413.

Lima-de Faria (1986) *Molecular Evolution and Organization of the Chromosome*. Elsevier Science Publishers, Amsterdam.

Löve A, Evenson V (1967) *Taxon* 16:423.

Marks GE, Schweizer D (1974) *Chromosoma* 44:405.

Mather K (1994) *Proc Roy Soc London B* 132:308.

McClintock B (1941) *Genetics* 26:552.

McClintock B (1950) *Proc Natl Acad Sci USA* 36:344.

McClintock B (1952) *Cold Spring Harbor Symp Quant Biol* 16:13.

McClintock B (1956) *Cold Spring Harbor Symp Quant Biol* 21:197.

Millard A, Spencer D (1974) *Aust J Plant Physiol* 1:331.

Miller OJ, Miller DA, Warburton D (1974) *Prog Gen Genet* 9:1.

Moffat AS, (2000) *Science* 289:1455.

Moore G (1995) *Curr Opin Genet Develop* 5:717.

Mukai Y (1996) In PP Jauhar (Ed.) *Methods of Genome Analysis in Plants*, 181 pp. CRC Press, Boca Raton, Florida.

Mukai Y, Nakahara Y, Yamamoto M (1993) *Genome* 36:489.

Mukherjee S, Sharma AK (1993) *Cytobios* 75:33.

Müller HJ (1927) *Proc Natl Acad Sci USA* 14:714.

Müller-Neuman MJ, Yoder P, Starlinger P (1984) *Mol Gen Genet* 198:19.

Müntzing A (1977) *Annu Rev Genet* 8:43.

Murray A (1991) *Current Opin Cell Biology* 6:872.

Nagl W (1975) *Prog Bot* 37:186.

Nagl W (1976a) *Nature* 261:614.

Nagl W (1976b) *Zellkon Med Zellkylen*. Ulmer, Stuttgart.

Nagl W (1976c) *Annu Rev Pl Physiol* 27:39.

Nagl W (1978) *Endopolyploidy and Polyteny in Differentiation and Evolution*, North Holland, Amsterdam.

Noll M (1974) *Nucleic Acids Res* 1:1573.

Nordenskiold H (1957) *Hereditas* 42:7.

Oehlkers F (1943) *Z Induktive Abstammungs-u Vererbungslehre* 81:313.

Olins A, Olins D (1974) *Science* 183:330.

Parasnis AS, Ramakrishna W, Chowdari KV, Gupta VS, Ranjekar PK (1999) *Theor Appl Genet* 99:1047.

Pardue ML, DeBaryshe PG (1999) *Chromosoma* 108:73.

Pardue ML, Gall HG (1970) *Science* 168:1356.

Parenti R, Guille E, Grisvard J, Durante M, Giorgi L, Buiatti M (1973) *Nature New Biol* 246:237.

Peschke VM, Phillips RL (1992) *Adv Genet* 30:41.

Peterson PA (1993) *Adv Agrom* 51:79.

Pelc SR (1972) *Int Rev Cytol* 32:327.

Pohlman RF, Fedroff NV, Messing J (1984) *Activat Cell* 37:635.

Prescott DM (1970) *Adv Cell Biol* 1:57.

Price HJ (1976) *Bot Rev* 42:27.

Ranjekar PK, Lafontaine JG, Pattota D (1974) *Chromosoma* 48:427.

Rao CSP (1956) *Annu Bot* 20:211.

Rayburn AL, Gill BS (1985) *J Hered* 76:78.

Renz M, Nehls P, Hozier J (1977) *Proc Natl Acad Sci USA* 74:1879.

Rick CM, Burton DW (1954) *Genetics* 39:640.

Rill RL (1979) In JH Taylor (Ed.) *Molecular Genetics* 3, 247 p.. Academic Press, New York.

Rogers J (1988) *Plant Mol Biol* 11:125.

Roy RP (1974) *J Ind Bot Soc* 53:141.

Ruderman JK, Baglioni C, Gross PR (1974) *Nature* 274:36.

Salisbury I (1995) *Current Biology* 7:39

Sandler SJ, Stayton M, Townsend JA, Raiston ML, Bedbrook JR, Dunsmuir P (1988) *Plant Mol Biol* 11:301.

Sarathi G, Mohan Ram HY (1974) *Experientia* 35:333.

Sawin K, Mitchison T (1990) *Nature* 345:22.

Scherwood SW, Patton JL (1982) *Chromosoma* 85:163.

Schwarzacher T, Leitch AR, Bennett MD, Heslop-Harrison JS (1989) *Ann Bot* 64:315.

Sears ER (1956) *Brookhaven Symp Biol* 9:1.

Sears ER (1981) In T Evan, NJ Packard (Eds.), *Wheat Science – Today and Tomorrow*, 75 pp. Columbia Univ Press, New York.

Sen H, Moore PH, Heinz D, Kato S, Ohmido N, Fukui K (1999) *Plant Mol Biol* 37:1165.

Sen S (1965) *Nucleus* 8:62.

Sen S (1974a) *Caryologia* 27:7.

Sen S (1974b) *Nucleus* 17:40.

Sen S (1978) *Experientia* 34:724.

Sharma A (1984) *Environmental Chemical Mutagenesis*, Prespective Rep Series No 6 Golden Jubilee Publication Indian National Science Academys, New Delhi.

Sharma AK (1974) In H Busch (Ed.) *The Cell Nucleus*, 263 pp. Academic Press, San Francisco.

Sharma AK (1975a) *Nucleus* 18:93.

Sharma AK (1975b) *J Ind Bot Soc* 54:1.

Sharma AK (1976) *Proc Indian Natl Sci Acad (Silver Jubilee Mem Lecture)* B42:12.

Sharma AK (1978) *J Ind Acad Sci* 87B:161.

Sharma AK (1979) In *Tropical Botany*, 327 pp. Academic Press, New York.

Sharma AK (1983) In PE Brandham, MD Bennett (Eds.), *Kew Chromosome Conference II*, 35 pp. George Allen, London.

Sharma AK (1984a) In VL Chopra, BC Chopra, RP Sharma, HC Bansal (Eds.), *Genetics: New Frontiers 1* 205 pp. Oxford & IBH, New Delhi.

Sharma AK (1984b) In Sharma AK, Sharma A (Eds.), *Chromosome in Evaluation of Eukaryotic Groups*, Vol II, 169 pp. CRC Press, Boca Raton, Florida.

Sharma AK (1985) Chromosome structure. *Perspective Report Series No 14*, p. 1. Indian National Science Academy, New Delhi.

Sharma AK (1999) *Meth Cell Sci* 21:73.

Shama AK and Aiyangar HR (1961) *chromosoma* 12:310.

Sharma AK, Gupta A (1959) *Nature* 184:1821.

Sharma AK, Mookerjea A (1954) *Bull Bot Soc GC Bose Mem* 8:24.

Sharma AK, Roy M (1956) *La Cellule* 58:109.

Sharma AK, Sharma A (1959) *Bot Rev* 25:514.

Sharma AK, Sharma A (1960) *Int Rev Cytol* 10:101.

Sharma AK, Sharma A (1980) *Chromosome Techniques: Theory and Practice*. Butterworths, London.

Sharma AK, Sharma A (1994) *Chromosome Techniques – A Manual*. Harwood Academic, Chur, Switzerland.

Sharma AK, Sharma A (1999) *Plant Chromosome: Analysis, Manipulation and Engineering*. Harwood Academic, Chur, Switzerland.

Shepherd NS, Schwart-Sommer Z, Velspalve JB, Gupta, Wienand U, Saedler H (1984) *Nature* 307:185.

Shui H, Miao Xing, Jian Zhao, Ming-da J (1992) *Plant Chromosome Research*, pp. 73–80, Proc Soc Sini-Jpn Symposium Pl Chromosome.

Shumny VK, Vershinin AV (1989) In AK Sharma, Sharma A (Eds.), *Advances in Cell and Chromosome Research*, 47 pp. Oxford & IBH, New Delhi.

Slijepevic P (1998) *Chromosoma* 107:136.

Smith AJE (1983) In AK Sharma, Sharma A (Eds.) *Chromosomes in Evaluation of Eukaryotic Group*, Vol. 1, 1 pp. CRC Press, Boca Raton, Florida.

Smith APJ (1978) *Adv Bot Res* 6:195.

Smith BW (1967) *Amer J Bot* 54:654.

Smith DR (1991) *Chromosoma* 100:355.

Stadler LJ (1928) *Annl Rec* 41:97.

Stebbins GL (1971) *Chromosomal Evolution in Higher Plants.* Arnold, London.

Stebbins GL Jr (1950) *Variation and Evolution in Plants No. XVI,* Columbia Biological Series, Columbia University.

Stockert JC, Lisanti JA (1972) *Chromosoma* 37:117.

Stubblefield E, Wray W (1971) *Chromosoma,* 32:262.

Svaren J, Horz W (1996) *Curr Opin Genet Dev* 6:164.

Swaminathan MS, Natarajan AT (1957) *Stain Technol* 32:43.

Sybenga J (1995) *Euphytica* 83:53.

Sugimura T, Kondo S, Takebe H (Eds.), (1982) *Environmental Mutagens and Carcinogens.* University of Tokyo Press and Alan R Liss, New York.

Sunkel CE, Coelho PA (1995) *Curr Opin Genet Dev* 5:756.

Swift H (1950) *Physiol Zool* 23:169.

Taylor JH (1984) In VL Chopra, BC Joshi, RP Sharma, HC Bansal (Eds.), *Genetics New Frontiers Proc XV Int Cong Genetics,* Vol. 1, 213 pp. Oxford & IBH, New Delhi.

Taylor JH, Woods PS, Hughes WL (1957) *Proc Natl Acad Sci USA* 43:122.

Vanderlyn L (1949) *Bot Rev* 15:507.

Van der Krol, Lenting PE, Veenstra J, Van der Meer IM, Koes RE (1998) *Gerata AGM, Mol* 9:217.

Venetski UP (1986) *Selskohoziastvenaya Biologia* 4:23.

Verma SK, Kumar SS (1980) *J Hattori Bot Lab* 47:245.

Vershinin AV, Heslop-Harrison JS (1996) *Plant Mol Biol* 36:149.

Vosa CG (1973) *Chromosoma* 43:269.

Vosa CG (1977) In K Jones, Brandham PE (Ed.), *Current Chromosome Research 105.* Elsevier, Amsterdam.

Warmke HE (1946) *Amer J Mot* 33:648.

Warmke HE, Blakeslee A F (1940) *Amer J Bot* 27:751.

Weising K, Nybom H, Wolff K, Meyer W (1995) *DNA Fingerprinting in Plant and Fungi.* CRC Press, Boca Raton, Florida.

Westgaard M (1958) *Adv Genet* 9:217.

Westgaard M (1940) *Dansk bot Arkv* 10:1.

Widom J (1997) *Science* 278:1899.

Williams JG, Kubelik AR, Livak KJ, Rafalski JA, Tingey SV (1990) *Nucl Acids Res* 18:6531.

Winge O (1923) *C R Trav Carlsberg* 15:1.

Yahasigawa T, Tano S, Fukui K, Harada K (1993) *Jpn J Genet* 68:119.

Subject Index

Author Index